P9-DMZ-047

FOOD ^{AND}_{THE} CITY

WITHDRAWN

Urban
Agriculture
and the
New Food
Revolution

FOOD AND CITY
THE

JENNIFER COCKRALL-KING

Prometheus Books

59 John Glenn Drive
Amherst, New York 14228–2119

Urbandale Public Library
3520 86th Street
Urbandale, IA 50322-4056

Published 2012 by Prometheus Books

Food and the City: Urban Agriculture and the New Food Revolution. Copyright © 2012 by Jennifer Cockrall-King. All rights reserved. No part of this publication may be reproduced, stored in a retrieval system, or transmitted in any form or by any means, digital, electronic, mechanical, photocopying, recording, or otherwise, or conveyed via the Internet or a website without prior written permission of the publisher, except in the case of brief quotations embodied in critical articles and reviews.

Trademarks: In an effort to acknowledge trademarked names of products mentioned in this work, we have placed ® or ™ after the product name in the first instance of its use in each chapter. Subsequent mentions of the name within a given chapter appear without the symbol.

Cover photo reproduced by permission from Compassionate Eye/Steven Errico/Getty Images
Cover design by Nicole Sommer-Lecht

Inquiries should be addressed to
Prometheus Books
59 John Glenn Drive
Amherst, New York 14228–2119
VOICE: 716–691–0133
FAX: 716–691–0137
WWW.PROMETHEUSBOOKS.COM

16 15 14 13 12   5 4 3 2 1

Library of Congress Cataloging-in-Publication Data

Cockrall-King, Jennifer, 1971–
    Food and the city : urban agriculture and the new food revolution / by Jennifer
Cockrall-King.
        p. cm.
    Includes bibliographical references and index.
    ISBN 978–1–61614–458–6 (pbk. : alk. paper)
    ISBN 978–1–61614–459–3 (ebook)
    1. Urban agriculture. 2. Food supply—Social aspects. I. Title.

S494.5.U72C63 2011
363.8—dc23
                                                                        2011041554

Printed in the United States of America on acid-free paper

For my parents, food lovers and gardeners extraordinaire

CONTENTS

INTRODUCTION

The idea to write a book about urban agriculture—the practice of producing and distributing food right in cities—felt like it came looking for me as much as I went looking for it.

As a food writer with a serious passion for gardening, I had long been in the habit of stopping to talk with anyone watering a few pots of rosemary and basil, for instance, on the patio. (Several minutes later, we'd still be trading stories about what interesting edibles could be grown with the right amount of obsessive coddling.) But about five years ago, I started noticing more tomatoes and cucumber vines twisting around condo balcony railings where previously there had only been the usual flowerpot standards of geraniums and lobelia. Then a few maverick homeowners began ripping up their front lawns and replacing them with tidy rows of pole beans, peas, and carrots. Other urbanites were not so subtly defying city bylaws and keeping chickens and beehives in backyards. Finally, it was impossible to ignore how community gardens continued to mushroom in size and quantity, not just in my hometown, but in other cities I visited.

Before long, I was obsessed with finding food growing in cities in unexpected places and ways. Moreover, I wanted to delve into the question of why there seemed to be a growing number of urbanites suddenly investing time and energy in growing, producing, sourcing, and supplying food much closer to home. I needed to know what was fueling this enthusiasm for growing vegetables, fruit, and even keeping a few laying hens or beehives in the city.

Were these simply the same diehard farmers' market-goers who were now spending a few bucks on seeds and a couple of hours a week in the garden? There was no denying that the farmers' market renaissance in the past decade had given many people a reminder (or in some cases, an education) of what fresh, seasonal food looked and tasted like. Was this all for the sheer pleasure of a couple of homegrown harvests without worrying what pesticides had been sprayed on the delicate baby eggplant, strawberries, and such?

Or was this merely a fleeting trend masquerading as a generational rediscovery of the domestic arts, like knitting and quilting?

I had a hunch that many people, like myself, were turning to growing their own food because their excitement had waned for ever-cheaper food, produced farther and farther away. Food scares had been shaking the tree of industrially produced cheap food in our free-market economy for over a decade, and it was becoming impossible to turn a blind eye to the risks we were opening ourselves up to in order to shave a few cents off our grocery bills. Food companies had become so big and their distribution lines so global, a contaminant scare could affect millions of consumers on an international scale. In 2003, bovine spongiform encephalopathy (BSE), also known as mad cow disease, was confirmed in a cow on a Canadian farm, shutting down cross-border trade in cattle and destabilizing beef prices on both sides for years. Also in 2003, a line of organic pear juice for children, produced in China, had to be recalled from shelves in the United States, Canada, and Hong Kong because of dangerously high levels of arsenic in the juice. A deadly strain of *E. coli* OH57:H7, a dangerous bacterium in fresh spinach packs from California, affected twenty-six states, sickened hundreds, and left five people dead in 2006. Those were just the beginnings of a decade of mounting food scares. It was becoming impossible to ignore the risks of out-sourced large-scale food production in the name of convenience and price.

Like many in the so-called food movement, I had pored over books such as Eric Schlosser's *Fast Food Nation: The Dark Side of the All-American Meal*, Marion Nestle's *Food Politics: How the Food Industry Influences Nutrition and Health*, and Michael Pollan's *The Omnivore's Dilemma: A Natural History of Four Meals*. (By taking on the ethical, nutritional, and safety questions of industrially produced food, Schlosser, Nestle, and Pollan finally got a critical mass of people asking questions about where our food was coming from and how it was being produced.) In my local Slow Food chapter, I counted myself among the new wave of conscious eaters who were watching their food miles—the distance a food travels between the farm to our forks and the amount of fossil fuels used in the journey. With the average grocery store item traveling more than 1,500 miles (2,400 kilometers) from the field to our plates, it seemed that we had completely lost control of the complex algorithm that was our food chain.[1] Concerns over the ethics of our cheap food and the politics of unequal access to fresh, healthy food were finally becoming mainstream topics of conversation.

I was also coming to the realization that counting my foods' food miles was a luxury that others, even in my own city, didn't have. Many large urban centers were coming to be known as food deserts because of the total absence of proper grocery stores, which therefore resulted in less access to affordable, nutritious, fresh food in their immediate vicinity. Usually what remained were convenience stores and fast-food outlets.

Was it any wonder that some of us were trying to find alternatives to the chemical fertilizer–laden, pesticide-dependent, fossil fuel–guzzling industrial food? It got me thinking about alternatives to what I had previously taken as a given, that food is grown and produced elsewhere, and if we are lucky, we buy it exclusively from a supermarket. I began to wonder how much food and what kinds could actually be produced in cities.

My interest in urban agriculture led me to Cuba in 2007. I took part in an island-wide "food tour" organized by Wendy Holm, a Canadian agricultural expert. Our group visited a number of Cuba's ubiquitous *organopónicos*—commercial urban organic farms across the island. These astonishingly productive food gardens, often growing in little more than a foot of soil in raised beds over concrete, opened my eyes to just how much food could be produced in a city when people really put their minds to it. I saw urban garden after urban garden bursting with gorgeous salad crops, vegetables, herbs, and even "green" medicines like the noni fruit, a pinecone-shaped knobbly tropical fruit high in vitamin C and sold in North American health food stores as the next "super-fruit." Moreover, Cuba's urban food gardens were not just in the capital of Havana but in every medium and small town our group visited from one end of the island to the other. Cuba, as it happened, had suddenly been forced to reinvent its national food supply. (As I explain in chapter 14, Cuba went through a sudden fuel crisis in the early 1990s when its dominant trading partner, the Soviet Union, dissolved. Cut off from any outside trading partners, the fuel crisis quickly became a food crisis, and Cuba had to reboot its food system very quickly, without any chemical inputs or fossil fuels—or starve.) As a result of this crisis, urban agriculture wasn't a fringe activity in Cuba but rather the keystone of their new, postindustrial food system. It became the island's low-tech but effective and sustainable model of how to grow and distribute fresh, nutritious food right in cities.

Around the same time, I was becoming increasingly uneasy about the vulnerability of the monolithic industrial food landscape we all relied on back at home. Grocery stores, from which we in the industrialized world get nearly all our food, operate on highly efficient, just-in-time, long-distance supply chains. The reality, though, was that these systems had become so efficient that cities had little more than a three-day supply of food on hand at any given time. If those supply chains were to grind to a halt due to a fuel crisis, border closures, war, or environmental catastrophes, the grocery store shelves would be empty within days.

Urban agriculture, food grown and sold right down the street, was suddenly starting to make a lot more sense, not just in Cuba but in cities closer to home.

Though my eyes were open to radical new possibilities for alternatives to the weekly trip down the grocery store aisles, I was not naive enough to imagine that cities in North America would consider giving up valuable urban land for such a marginally profitable enterprise as growing food. As a gardener, I also knew that nonindustrial agriculture requires a human physical effort that most of us are no longer fit for. We were so far from even acknowledging the intrinsic problems in our food system at home that a food revolution, in my mind, was a very long way off.

That said, I was encouraged by the gathering momentum to challenge the status quo of what kind of food we had access to and where it was grown, even in my hometown of Edmonton, Alberta, a northern Canadian city under snow for six months a year. Looking back, it was on a bitterly cold evening in November 2008 that I started to consider that real change was on the horizon for how food and cities could move forward together.

That evening, I made my way to a public hearing at city hall, expecting to be part of a small group of concerned citizens voicing concerns about the city's thirty-year development plan. The document addressed housing and transportation plans to cope with the projected population and economic growth targets up through 2040, not exactly the stuff that tends to get people to leave their couches and televisions on a November weeknight for the municipal city council chambers' public gallery. Yet it touched a nerve, as the plan failed to protect the last of the city's development-threatened urban farmland, prime agricultural land where a few remaining market gardeners still produced an astonishing array of fresh produce for local markets despite a short growing season of just over one hundred frost-free days. Nor was there mention of a municipal food policy—that is, a deliberate public policy plan for a safer, healthier, more egalitarian approach to the production, distribution, and consumption of food in the region—despite a grassroots civic lobby that looked like it was finally getting some traction on the idea. Instead, the thirty-year municipal development plan was likened to a blueprint for a massive home renovation that curiously did away with the kitchen.

I arrived at city hall that evening to be shuttled into the third over-flow room that had to be quickly opened. More than 550 people had shown up, not just a few food activists and weather-beaten farmers waxing nostalgic for a lost way of life. Edmonton city councilors had never seen anything like it. I venture to guess that neither had the orga-nizers, the Greater Edmonton Alliance, a citizen action group that was only a few years old at the time.

The crowd featured a range of ages, ethnic groups, incomes, and other social groups that rarely met under one roof. In compliance with the rules of a public hearing, representatives from several of the citizen groups were allowed to address the city councilors, and the entire evening was piped into all the overflow rooms through closed-circuit audio. Restaurateurs spoke about the importance of preserving local foods to ensure an authentic local food culture. Parents voiced their safety concerns with the global food chain that begins thousands of miles away. Church leaders spoke about moral duties of land steward-ship, and community leaders spoke about the need to shorten the food chain, to buffer against food shortages and price increases, as well as to involve community members in creating their own resilient food safety net. In short, food—for the first time in a few generations—was back on the political menu. The fact that hundreds of people came out in sup-port of shortening the food chain, taking the city's food security into their own hands, and showing their support for growing food in the city was beyond significant and impossible for the city councilors to ignore.

If this was happening in my city, I wagered that seeds of change were germinating in others as well. Indeed, a few months prior, a group of Slow Food activists had caused a sensation when they planted what they called a "Victory Garden" in front of San Francisco's city hall that yielded more than a hundred pounds (forty-five kilograms) of fresh produce for the city's food bank. By the spring of 2009, First Lady Michelle Obama would dig up of a portion of the South Lawn at the White House and plant 1,100 square feet (102 square meters) of organic vegetables. And the Seattle city council would declare 2010 its official year of urban agriculture.

Seemingly everywhere, cities were forming food-policy councils; community gardens were multiplying; and municipal governments were voting on whether to allow households to keep a few urban chickens or a beehive, or to permit commercial farming to coexist with other commercial pursuits in their cities.

Across the pond, I learned that London had thirty thousand allotment gardens, as long-lease community gardens are called in the United Kingdom. There was a municipally supported push to create another 2,012 food gardens in the city by the opening of the 2012 London Olympic Games. City farms—small mixed farms in the middle of major cities, many of which include cows, chickens, ducks, and goats as food livestock—were a booming businesses all over the United Kingdom, and community orchards were now catching on as fast as are community gardening clubs. And Paris, which is the birthplace of modern intensive urban agriculture, was experiencing its own revival of urban gardening. Parisians were also keeping urban bees in record numbers.

Looking back, revolution was in the air. The domino effect of the major global economic meltdown that began in 2007 quickly turned into a series of food shocks—sudden, dramatic price increases for certain foods or the complete loss of supply of them—in various countries almost simultaneously. Though caused by years of the financial industry's overreaching in the big industrialized countries of the United States and the United Kingdom, the immediate consequence of the global credit crisis played out more concretely in the streets and in the food markets of Egypt, Yemen, Russia, and India. The prices of basic food staples like rice, wheat, and potatoes soared, causing riots. Soon afterward, Mexico experienced its own surge in the cost of corn, leading to a series of "tortilla riots."[2] Italy endured a "pasta strike" when outraged consumers refused to accept the 20 percent cost increase of pasta.[3] Argentineans demanded that their government stabilize the price of tomatoes as they shot out of reach of ordinary citizens.[4] These staples, which by now were commodities bought and sold on the open world market, like anything else, were subject to the price volatility of a nervous global market. Furthermore, because of the fact that staple products like rice, corn, and wheat were now considered inputs in the burgeoning biofuels industry, the price of food staples was now tied to the cost of fossil fuels.

Industrial food, for all its decades of bluster about feeding the world's poor, eliminating starvation, and ending global hunger, simply supersized the problems. By squeezing out record crop yields, we exhausted and eroded the soil as we increased our planet's population, which tipped over into seven billion in 2011 and is projected to be nine billion by 2045.[5]

The problem is that in that time, we've failed to solve the inequality problems of who overeats and who is malnourished. Now we are hearing those same arguments from those who tout genetically modified foods and crops as the solution to world hunger. They say genetic modification is necessary to feed the planet, and they use the same logic to push their for-profit patented foods, not just in selected "needy" countries but globally as well.

Closer to home, hunger soared as the middle class became the working poor, and the poor became destitute.

Food security, as defined in 1996 at the United Nations' World Food Summit, exists "when all people at all times have access to sufficient, safe, nutritious food to maintain a healthy and active life."[6] This often refers to both a physical and an economic ability to obtain good food. When one or both of these conditions are not met, individuals, groups of people, and entire nations can be said to be food insecure.

In 2008 and 2009, 50.2 million Americans were food insecure—that is, they didn't know where their next meal was coming from on any given day.[7] This figure included the 17.2 million children in those households.[8] The US Department of Agriculture's Economic Research Service, the group that has tracked food-security issues in the United States since 1995, reported that these levels were the "highest percentage observed since nationally representative food-security surveys began in 1995."[9] The report also noted that "food insecurity was more common in large cities than in rural areas and in suburbs and other outlying areas around large cities."[10] There was a rise in food deserts, areas in cities devoid of markets and grocery stores, where people have little or no access to healthy, nutritious, fresh, whole foods because grocery stores moved out to the suburbs, where more affluent customers had also moved. Low-income communities in the city centers were left with fast-food outlets and convenience stores as their only options. Without easily accessible and affordable transportation options—for the past decade, the average distance Americans must travel between their home and the closest grocery store is six miles—good food ended up out of reach physically as well as economically. (Eventually, people who live in these food deserts forget or fail to learn how to recognize whole foods and how to prepare a meal from scratch, leading to a kind of food illiteracy.)

Clearly, the urban-agriculture movement wasn't happening in a vacuum. The more I learned about the desperate situation that we were in as industrial consumers, the more I grew to appreciate how revolutionary, subversive, and necessary the open-source, chaotic, decentralized nature of the urban-agriculture revolution seemed. If the pundits' predictions of a catastrophic failure of a century-long experiment in an industrialized and, more recently, globalized food system ever came to pass, community gardens, urban chickens, public orchards, urban beekeeping, commercial urban farms, open sharing of knowledge, and even the science fiction–like promise of vertical farms were poised to coalesce into a new urban food revolution. A shorter food chain, as any economist would tell you, was the future.

Once I started following the thread of urban agriculture, it was all I could do to keep up. I was bombarded with e-mail updates from my friends with news about their urban chickens, complete with glamour photos of laying hens and boasts about egg counts. My friend Patty Milligan reported that she couldn't keep up with requests for her beginner urban beekeeper courses, and Jeremy, my thirteen-year-old neighbor, finally convinced his parents that his fascination with bees and honey wasn't just a phase. (His two wooden beekeeping hive boxes, which he skillfully hand-painted with cartoon bees and flowers, are thriving and producing lavender and mint-scented urban honey.) And I watched how, within a few short years, it seemed as if the social event of the year—for urban hipsters, community activists, and baby boomers alike—had become the springtime seed exchange in community leagues all over North America.

As I decided to write a book about this flurry of activity in urban agriculture, I knew I had to include a wider perspective beyond my own city.

Heading out on the road, I traveled to cities at the forefront of taking back their food systems and adapting them to an urban environment. I sought out the forward-thinking pioneers who had been working in the urban-agriculture movement, as well as the fresh-faced newcomers oozing enthusiasm.

I spent time in well-known hotspots of urban agriculture in North America, such as Vancouver, British Columbia, where 42 percent of households grow food in their yards at home,[11] and Toronto, Ontario, where 40 percent live in households that produce some of their own food.[12] I was fascinated by the game-changing projects taking root in unlikely places like Milwaukee, Detroit, and Chicago. I ventured as far away as I could, to London where I met Mark Ridsdill Smith, who grows thousands of dollars worth of fresh vegetables, fruits, and herbs on his six-by-nine-foot balcony and in a few window boxes of his flat. In turn, he introduced me to Azul-Valérie Thomé, whose Food from the Sky is essentially a farm on the rooftop of a grocery store, situated so that vegetables can be picked, packed, and purchased within just a few hours from dirt to display cooler. I obsessed over the burgeoning urban wine movement in London, only to find that urban vineyards were everywhere in Paris (132 in the greater Paris region) and as far away as Vienna (1,700 acres [687 hectares] under vine within the city limits). I also added to my collection of urban beekeeping success stories when I found a thriving urban beekeeping community in Paris helped along by a decade-long ban on chemical pesticide use in the city limits.

And then I returned, once again, to Cuba, which still leads the global pack in permaculture food production—ecosystems modeled on the ones found in nature with high levels of biodiversity—a type of managed, edible wilderness that represents the latest thinking on a post-agricultural food system. No wonder Cuba has the world knocking on its doors to learn and exchange knowledge.

Revolutions happen for a reason, and I do my best to explain this in chapters 1 through 5. Two-thirds of Americans are overweight or obese, and yet, paradoxically, they are undernourished at the same time. Hunger is a growing problem not just in faraway places but right in our hometowns, with demands on the services of food banks and other hunger-relief programs reportedly increasing by double digits annually. Our cities are just microcosms of our entire planet—one billion people are overfed while another billion people go to bed hungry every night— where we have more than enough food, but those in need get too little or the wrong kind of food. Finally, our food crisis has created a health crisis that is threatening to overwhelm the budgets of every industrialized nation's government.

We need to understand how we got to the point where our food system has become so unhealthy, so unfair, so environmentally destructive, and has become a catastrophic failure. And just as the optimists and deniers kept repeating that banks were "too big to fail," our globalized, industrial food system has seemed to reach that point as well. If it's toppling under its own greed and mismanagement, it is because we allowed it to get too big to feed us properly. If we can no longer outwit the cold, hard realities of peak oil, peak water, and peak soil, we're in for one hell of a market correction. As we move into an uncertain economic future, food security will continue to erode in many communities that are already struggling.

That is why I have devoted the bulk of this book to telling the stories of people keeping a few chickens in their backyards (sometimes illegally, as in Toronto), farms being planted high on rooftops in the concrete jungles of cities (New York City and London), public orchard projects in unexpected places (Calgary, Alberta, Canada), community gardens being used as tools of social change (Vancouver's Intercultural Community Gardens project), and an entire country that has embraced urban agriculture as the cornerstone of its national food system (Cuba). The bulk of this book, chapters 6 through 14, therefore, is about the people, cities, and urban gardens and farms that are the seeds of change in this new urban food revolution.

In the five years from when this book began to take shape to its publication, the idea of food production within urban landscapes has gone from a fringe concern of a few academics, green thumbs, and counter-culture gadflies to a real hands-in-the-dirt mainstream, social, environmental, and economic revolution.

My travels, interviews, and discoveries did not always follow a linear path. But I've done my best to recount them in a way that makes geographic sense or at least thematic sense. If I meander and weave somewhat in the telling of these stories, that's just the gardener in me. You'll find very few straight rows in my garden.

Up to the very moment this book went to print, I squeezed ever-more stories onto the pages out of sheer admiration for ground-breaking projects that appeared daily on my radar. The London Olympic Games in the summer of 2012 will be the first major sporting event with a food policy in place, specifically the sourcing of food from the projected 2,012 urban food gardens in the city. The race is on to build the world's first vertical farm, a layered mixed farm—offering fish, fruit, veggies, and herbs—in the middle of a city. And plans for futuristic "agro-parks," which are spooky factory farms with a tourism component, continue to emerge from China, though it's unclear who exactly will build these and when.

The momentum behind urban agriculture as it stands in 2012 leaves me hopeful that this is a major turning point for how we design and use our urban spaces, how we feed ourselves, and how we treat our food producers and our planet. I am more convinced now than ever before that this is more than just a flash-in-the-pan green trend, and that the movement is showing no signs of slowing. As Paul Hughes, the highly quotable local foods and urban chicken activist in Calgary, Alberta, likes to remind people: "This is just the tip of the iceberg lettuce."

Chapter 1

THE FACADE OF THE MODERN GROCERY STORE

Pile 'em high, sell 'em cheap.

—Business motto of Jack Cohen,
who in 1919 founded Tesco,
currently the United Kingdom's
largest supermarket chain

THE SUPERMARKET

No matter how philosophically "locavore" (trying to source only locally grown and raised food), or how pro-farmers' markets I am, I still find myself pushing a shopping cart up and down the aisles of a supermarket a couple times a week.

I'm remarkably typical, as it turns out. According to the Food Marketing Institute, an Arlington, Virginia–based grocery retail association representing three-quarters of all retail grocery sales in the United States, the average supermarket shopper makes 1.7 trips to the supermarket per week.[1]

First of all, supermarkets are undeniably convenient. They are generally open seven days a week, and some chains and locations stay open twenty-four hours a day. I know what's available even before I get there, and most of the food is cheap. That said, if I'm willing to splurge a bit, I can also get strawberries in January. No wonder this is how the majority of us get our groceries.

The Rise of the Supermarket

For such a monopolistic hold on our food dollars, you'd think that supermarkets had been around since the dawn of time. It's difficult for us to conceive of how it could be otherwise, so it's shocking to think that they've been around for barely four generations. Academic, food-justice activist, and writer Raj Patel points out in his book *Stuffed and Starved: The Hidden Battle for the World's Food System*, "[S]upermarkets are patented inventions, and like all innovations, they responded to a specific need at the time and place of their conception."[2]

At the turn of the twentieth century, the industrialized nations—and the United States in particular—got very good at producing things, including food. With tractors, combines, and other mechanized farming equipment rather than plow-horses and human labor as the limiting factor of their workday, farmers could clear, plant, and harvest a much larger area, working bigger farms than ever before. They specialized to maximize the usefulness of their equipment and turned from being producers of a wide variety of crops and livestock on a farm (mixed farming) to specializing in as little as one single grain, pulse, or oilseed (monocrop farms). Readily available chemical fertilizers, pesticides, and other chemical treatments to the soil enabled single-crop plantings, which otherwise would exhaust the soil within a few harvests, just as antibiotics allowed concentrated feedlots that would otherwise render livestock sick and unsuitable for slaughter and sale. Farms essentially became factories that specialized in the efficient production of a narrow range of products—but in large quantities at a low unit price.

And just as industrial processes enabled industrial agriculture, industrial agriculture produced industrial food, even more so when food manufacturers began to rely on a narrow selection of raw ingredients that could be endlessly recombined into packaged and prepared items with a long shelf life.

Food production became so efficient at the end of the industrial food chain that scarcity quickly became overproduction. As with other retail goods, it was noticed that dropping the price encouraged people to buy up the surplus, even when it wasn't really needed. (Just like when commodity crops overproduce, the emphasis turns not to reducing production but to increasing consumption.)

Yet to *really* increase consumption, food retailers needed to invent the modern self-service grocery store, which capitalized on both the new concept of a self-service model paired with new, cheap, industrially produced food to set the wheels in motion for the weapons of mass consumption that we North Americans have become.

Though it seems almost ridiculous to contemplate, acquiring groceries in the past meant that you'd give a shopkeeper a list of items and quantities you wished to purchase. (It's important to keep in mind that the general store was mainly a dry-goods store with a few other items like bananas, citrus, and maybe raisins available. Fresh produce was bought at a farmers' market–style central market. And fresh meat came from the butcher shops or from butchers' stalls at the city market.) The shopkeeper would then assemble your grocery order for you and hand it across the counter once you had paid. Or, the tally was added to your store account if you were in good standing, credit-wise, with the shop owner. Most items were fetched from back rooms. A shopkeeper or clerk might suggest you try a new product that just came in, but impulse buying was not the norm. Getting groceries in those days also meant a number of stops at different specialty stores.

In 1914, brothers Albert and Hugh Gerrard had an entrepreneurial idea to combat the steep rise in grocery prices due to the First World War. They came up with the concept of cutting overhead by letting customers choose their own groceries themselves, right off the shelves. To help people find the items they were looking for on their own—a radical new idea that was sure to cause no end of confusion—the Gerrards decided to assist customers by stocking the food items alphabetically. They called their California-based stores the Alpha Beta.

Another grocer, Clarence Saunders, had a similar idea, but he thought it through a bit more. In 1916, Saunders opened his first self-service King Piggly Wiggly grocery store in Memphis, Tennessee. This new Piggly Wiggly retail model had customers entering the grocery floor via turnstiles and carrying a shopping basket as they were set on a course that snaked up and down each aisle, with a single direction of traffic flow, until the customer reached the checkout and was released back out through an exit turnstile just past the checkout register.

(While we are allowed to roam more freely in today's supermarket, the basic mazelike design is still how most grocery stores are designed, with the added retail trick of placing staple items at the far reaches of the store, forcing us to cover as much geography and to pass as many higher-profit products as possible. Except for a few items such as cars, cosmetics, and perfume, the self-service retail model dominates most consumer goods shopping experiences.)

Saunders, not the Gerrards, was the first to file his idea at the patent office. In 1917, he received US Patent 1,242,872 for his concept for the "Self-Serving Store." In less than a decade, 1,200 Piggly Wiggly stores opened across the United States.[3] By 1932, there were 2,660 stores.[4]

Saunders also came up with the idea for the self-service checkout to fully automate the grocery experience in 1937. Only a few Keedoozle—an awkward combination of "key does all"—food stores were built, and it was clear that the automated vending technology just wasn't where it needed to be. The self-serve checkout, as anyone who shops at a supermarket now knows, is finally a reality in the retail landscape, fifty years after Saunders's failed prototype stores. I'm amazed at how unconcerned we seem to be that checkout clerks, the only real human interaction left for the consumer in the industrial food chain, are being phased out. You don't see the farmers, the fishermen, the ranchers, or the fruit growers who produce your food. Soon, we will no longer see people who swipe it past the barcode scanner and process our payment. (For now, I remain defiant and queue up with the rest of the holdouts at the last few human-operated supermarket checkouts.)

Within a few generations, we have unquestioningly accepted the industrial food system—and the supermarket model serving as its retail outlet—that is concerned merely with lowering the unit costs of the food in question. To say that this is the dominant model is stating the obvious, given the fact that the industrial food system now provides us with 99 percent of the food we eat in North America.[5] In return for our loyalty to this model, we get 38,718 different food items to choose from in an average grocery store.[6] There are seventeen thousand new hope-fuls—new food products—launched into this grocery landscape every year. And talk about cheap! We Americans devote less of our income to purchasing food than any other nation, around 9 percent on average, which is less than what we spend for our transportation needs.[7]

So what's the downside?

THE ILLUSION OF CHOICE

Within the grocery store, we have the illusion of choice. Forty thousand items sounds like a lot of choice, but it's nothing compared to nature's inventory. The United Nation's Food and Agriculture Organization (FAO) estimates that in the twentieth century, 75 percent of the biological diversity of our foods has been lost as a result of industrialized agriculture. Other sources claim that we've lost up to 90 percent of our global food biodiversity.[8] The variety of food plants such as the different types of carrots, beans, or lettuces being grown; the genetic diversity of aquatic food stocks; the variety in breeds of our livestock animals; and the total biological diversity of the food that we draw on has been drastically reduced in our lifetime. Rather than having fewer food choices available than we do now, our grandparents actually had more.

Why? Diversity is the enemy of mechanization, so industrial agriculture values consistency, uniformity, and durability for long-distance transport, a quality like taste doesn't enter into the matter. Imagine what we have lost in the flavor pantheon that existed just a few generations ago. And since then, we've lost a full 97 percent of the varieties of our fruits and vegetables, thanks to the unnatural selection of the industrial food system.[9] We're losing about 2 percent of the genetic diversity of the world's crops per year.[10] We'd better get to learn to like a smaller and smaller selection of foods. Only 150 different food-plant species are grown on a large commercial agricultural scale in the world.[11] Despite the fact that farmers have domesticated over five thousand plant species, the industrial food chain uses a mere 3 percent of them.[12]

For example, there are hundreds of varieties of apples in North America. They come in different sizes, shapes, and colors. They all have slightly different coloring, textures, and flavors. Some store well; others don't. Some are best for baking; others are best for drying. Some make great apple juice; others make excellent cider. Some ripen on the trees in June; others must hang until October. Now *that* is choice.

Sadly, we don't get these choices at the grocery store. Last time I looked there were Granny Smith, Golden Delicious, Spartans, Fujis, and maybe Pink Lady® apples, if we were lucky. All are meant for eating raw, and they are all chosen for their ability to be picked while bullet-hard to ship without bruising and to store for months. There's nothing there for me if I want to make my own apple juice or bake a really good pie.

The same goes for tomatoes. Of the hundreds of types and shapes that exist, we're allowed only the ones that are tough enough to endure the industrial food chain.[13] If you want to taste a tomato picked only when it was ripe, you are out of luck at the supermarket. And the choice of these apples and tomatoes is alarmingly consistent throughout the calendar year. This uniformity of choice is what is known as global summertime: when grocery supply lines reach all the way around the globe, it's always summer somewhere. The produce selection in January is pretty damn similar to the produce selection in June, yet it shouldn't be. Broccoli actually is a seasonal product. Peppers are too. So are strawberries, apples, and tomatoes. You'd just never know it from the inside of a grocery store.

Moreover, the choices that we are presented with on the shelves or in the cooler aren't often true choices. We can choose between brands of eggs, but when a salmonella contamination scare on two Iowa farms in August 2010 resulted in a nationwide recall of a half billion eggs, dozens of different brand names of eggs were affected. Why? Because five hundred million eggs all originated from one large-scale producer. Sure, they were sold under different brand names, but they all came from the same huge corporate farm. If that's not sobering enough, consider that there are a mere five corporations behind 90 percent of the US food supply.[14]

And outside the supermarket, the illusion continues. We can choose between major chains, but in the end, we have very little choice of how we get our food other than via a supermarket. Choosing between Costco and Walmart is simply the choice between Coke® and Pepsi®. It's essentially the same stuff on the inside.

NINE MEALS FROM ANARCHY

A topic discussed in food-security circles—those groups of people who track food reserves that exist in a city or a country at any given time—that gets surprisingly little coverage in the general discussions about food is the estimate that cities nowadays have a mere three days' worth of food at any given time to feed their populations.

In 2000, farmers and transport truck drivers in the United Kingdom staged a protest over government fuel duties that they felt were crippling them, along with the rising cost of gasoline and diesel. Their strategic protests and blockades managed to severely disrupt the nation's fuel supplies, shutting down motorized transportation. There were also so-called rolling blockades on major highways to disrupt transportation in and out of cities. The major grocery chains, Sainsbury's, Tesco, and Safeway, noticed panic buying, and without reliable deliveries to restock their shelves, they started rationing their food supplies by the third day.

The British government took notice of how quickly a city like London could run out of food. It created an agency called the Countryside Agency to study the United Kingdom's food security. In 2007, Lord Cameron of Dillington, the head of the Countryside Agency, concluded in rather dramatic fashion that Britain was indeed extremely vulnerable to a food shock caused by any disruption of the normal flow of supply lines. Major cities in the United Kingdom, the report concluded, were at any given time "nine meals from anarchy."[15]

That the big supermarkets knew to start rationing what supplies they had left on the third day was not by coincidence. Though they likely don't refer to it as "nine meals from anarchy," they do operate what is known as the three-day rule.

The supermarket retail business is highly competitive. Supermarkets depend on volume to turn a profit, because they average less than 1 percent net profit after tax in a year, according to the Food Marketing Institute's published 2010 figures.[16] More to the point, they depend on tightly controlling their costs so as not to lose any of that profit. Holding a lot of inventory, in a grocery retailer's mind, is costly. Milk, bread, fresh fruit, and vegetables—pretty much any perishable inventory, so much of which is thrown out as it wilts, rots, passes its sell-by date, or goes moldy—is the worst kind of inventory for a grocery store. It's the loss leader that gets you in the store, but it's also why these products are tucked in the back of the store, forcing you to walk past the other nonperishable, more expensive, processed items.

To carry as little inventory as possible, grocery chains have created very sophisticated just-in-time "value chain logistics" systems. They manage inventory so well that they only need a three-day supply of food in their distribution system at any given time.

It's worth pointing out that these tightly controlled supply lines that replenish our grocery stores are fine—until they're not. When so much of our food comes from so far away, what happens when there is a disruption in fuel supplies, a natural disaster blocking access to a city, or a terrorist attack shutting down borders and internal transport? Three days of food is not enough inventory.

When the planes flew into the twin towers of the World Trade Center in New York City in 2001, transportation within the United States was severely restricted, but it eventually had to be reestablished because the city would soon run out of food after about three days. The same three-day rule played out in New Orleans when Hurricane Katrina left the city cut off from resupply. And this can be said for any city in the United States or Canada at any given time. Ask anyone with firsthand knowledge of the modern grocery "supply chain" system, or any emergency preparedness policy maker. They probably can't tell you on the record, but they'll be unable to deny that there are only a few days' worth of food in the city.

The supermarket is really just the outlet mall for the industrial food system. And if, as so many organizations and economists warn, we're coming to the end of the industrial food system, or if we're at least facing a major crisis in it, we'd better start figuring out how to feed ourselves when those shelves go empty for the very last time.

This is why, as I stood in the produce section one day despairing for the unglamorous, unloved, yet local and seasonal root vegetables and heads of cabbage, the illusion of abundance that is the supermarket model unraveled for me. I realized that it was all smoke and mirrors. We were, as Michael Pollan so astutely writes, "eating at the end of the industrial food chain" to our bodies' and our environments' detriment.[17] But what if we were also *coming to the end* of eating at the end of that industrial food chain?

Chapter 2

INDUSTRIAL FOOD

Any customer can have a car painted any color he wants as long as it is black.
—Industrialist and inventor of the
Ford Model T, Henry Ford, 1922

At the beginning of the 2008 documentary *Food, Inc.*, Michael Pollan explains the trade-off we've made for convenient, cheaper food at the till and for triple-washed, pre-cut, bagged lettuce. "The way we eat has changed more in the past fifty years than in the previous ten-thousand, but that's the image that is used to sell the food. . . . You go into the supermarket and you see pictures of farmers. The picket fence and the silo and the 1930s farmhouse and the green grass. The reality is . . . it's not a farm. It's a factory. That meat is being processed by huge multi-national corporations that have very little to do with ranches and farmers."[1]

These are not just the pronouncements of food activists, journalists, and filmmakers. Ask your grandparents, parents, aunts, uncles, or elderly neighbors. They'll tell you about the kinds of food that appeared on their table when they were kids and what grocery stores were like in their youth. They'll tell you about how food grew in gardens next to the house and how the bounty that didn't end up on the table that day was canned and stored in the basement for winter. Then they'll tell you that food looked nothing like the toaster pastries and SpaghettiOs® that we stock our cupboards with nowadays.

I certainly didn't need to scratch too deeply to find real-life examples of this change. My grandparents on both sides, for instance, grew up on farms and thrived on a regional, seasonal food system. What they couldn't make, grow, or raise themselves was purchased from the general store in a nearby city. The constant trips to the supermarket only became routine for my grandmothers later, as married women living in cities, with working-class husbands and hungry mouths to feed at home. It must have been exciting times as they parked their newly acquired cars and stepped into the state-of-the-art supermarkets in the early 1950s. Little did they know that they were stepping from one era into another.

For example, my grandmother on my father's side was born in 1911 in a little farmhouse in Saskatchewan. She married my grandfather at the age of twenty-two, and by the time their second child came along, they had left the farming life to seek their future in the city. Another child came along just before World War II, and then the last of four was born the year before the war ended. Those were very lean years. Seeing her four precious children deprived day after day, year after year broke her heart, she'd tell me decades later with tears welling up in her eyes. (Whether it was due to lack of money, food shortages, or the eventual food rationing, my father still bristles at the sight of a bowl of porridge, a legacy of eating it three times a day as a young child. And my grand-mother spent the rest of her life making sure no one left her table hungry ever again.)

Food shortages and high prices continued after the war. In Canada, sugar rationing lasted until 1947. The very day, however, that it was de-rationed, my grandmother went out and bought a bag of sugar. There wasn't any supper that night. There was cake. One cake for each child. (It's worth noting that North Americans got off rather easy. Rationing didn't end in Britain until 1954.)

In those days, no one thought twice about growing food in a small garden plot at home. Sometimes, entire backyards were used for food production. It's just something people did in times of crisis.

Even before World War I, vacant-lot gardening was an organized movement in most cities in North America and Europe as a type of pre-food bank for families and individuals who couldn't afford to buy food. During the war, the idea of urban food gardening went from a relief situation to a war effort with the Victory Garden movement. Because so many farmers went off to war, food production dropped dramatically. An army of civilian gardeners was created both as a morale booster on the home front and, perhaps more importantly, to mitigate food shortages and steep rises in food prices. The year World War I ended, one particularly enthusiastic study of Victory Gardens, *The War Garden Victorious* by Charles Lathrop Pack, put "the number of such gardens at 5,285,000" in the United States.[2]

After World War I, interest in urban food growing dipped temporarily, only to roar back again throughout the worst of the Depression with "relief gardens." The following decade brought World War II, and the Victory Garden frenzy began again. The United States Department of Agriculture released a twenty-minute public service film called *Victory Gardens*, which "conscripted" families to do their part for the war effort. It explained how to turn one-quarter of an acre next to the house into an early garden, a late garden, and "where the *real* garden began, the Victory Garden," for canning and preserving to tide the family over for the winter and maybe as an insurance policy for food shortages. There was even a special certificate that a family could post beside its Victory Garden. "Our family will grow... A VICTORY GARDEN... in 1942. Realizing the importance of reserve food supplies, we will produce and conserve food for home use," it read.[3] It was partly a plaque of recognition and part patriotic pledge. There was even a line for a Victory Gardener's signature at the bottom.

Growing your own Victory Garden was heavily promoted as being both stylish and patriotic. Posters and gardening guides extolling the virtues of being self-sufficient food-wise were widely distributed. In *Victory Backyard Gardens: Simple Rules for Growing Your Own Vegetables*, published in 1942, the introduction begins, "In this world-wide war in which every American plays a part, every productive vegetable gardener helps our national well-being and is an aid to victory."[4] First Lady Eleanor Roosevelt even joined the ranks of World War II Victory Gardeners by planting a food garden at the White House, although she was strongly warned against it by the secretary of agriculture, Claude Wickard, who suggested it would hurt national food processors' commercial interests.

The World War II mass mobilization of citizen farmers in the United States was an extremely successful campaign. At its peak, the United States was producing 40 percent of its own vegetables in twenty million home-scale Victory Gardens on both private and public land.[5] Similar widespread citizen farmer movements were taking place in other countries during wartime, on both sides of the fight.

Until her children were grown, and even beyond, my grandmother relied on her resourceful farm-bred ways even in the city. She grew a seasonal "kitchen-garden" in the backyard with the basics: potatoes, carrots, onions, cabbage, herbs, beans, peas, and so on. She pickled cucumbers, canned homemade sauerkraut, and put up preserves to last through the long winter. She canned peaches, pears, apples, and cherries; the selection of preserved summer fruits tucked away under the stairs became the pride of her winter pantry. In the late fall, the root vegetables in the garden were dug up and stored in sand in the basement's cold room. Just before winter settled in, my grandfather and his buddies went on their annual fall hunting expedition for duck, goose, and the occasional pheasant. That would last throughout the winter, and by spring, the freezer would fill with pickerel, whitefish, northern pike, and lake trout from nearby lakes. My grandparents were by no means self-sufficient, food-wise, but their grocery expenses were extremely modest when supplies and money allowed: coffee, tea, tinned ham, cocoa, flour, bananas, raisins, tinned goods, and a bright box of Velveeta®, North America's beloved processed cheese product, as a treat. Their food chain was mostly short, with a few imported and processed exceptions.

My grandparents' fortunes steadily improved during the postwar economic boom. General stores and "groceterias" gave way to department store food floors. National chain supermarkets took over from independent grocers, offering an ever-expanding selection of tinned foods and exotic fruits like fresh pineapples, avocados, artichokes, and iceberg lettuce. Nevertheless, meals were made from scratch and eaten in the home. Restaurants were for very special occasions.

Like most women who had spent more time than they could tally standing at the stove and kitchen sink, my grandmother welcomed the ever-expanding selection of prepared foods in the 1950s and 1960s. TV dinners, eaten on newly purchased TV trays in front of the television, were beyond exciting, modern, and glamorous in their own way. Less time spent peeling, chopping, and standing at a hot stove was fine by her, and more and more processed food crept in. Her home cooking saw some exciting advances like ambrosia salad, which consisted of tinned fruit cocktail, tinned mandarin orange segments, and miniature marshmallows bound together by Dream Whip® edible oil topping. Tinned pineapple was also dumped on everything—especially products like Spam® and its similar canned pork cousins Klik, Kam, and Prem—to make a dish "Hawaiian." Convenience was king. Live television commercials showing a groaning spread of fluorescent Jell-O® moulds and Miracle Whip®–based foods. Home economists and cookbook authors echoed the intoxicating liberation of Minute Rice® and instant puddings that came with the modern lifestyle.

I can just see my grandmother quietly cheering the change in attitudes toward her domestic duties as wife and mother. "There is no virtue in doing things the hard way," wrote Canadian newspaper food section columnist Muriel Wilson in her *Victoria Times-Colonist Cookbook*. "It's fun getting acquainted with the newest mixes, the fabulous instants, the ready-to-go canned and frozen foods. Our jet-aged cooking would have astounded Grandma who, when she wanted gelatin, had to boil a pot of calves' feet for hours. Now we simply reach for a package. A package plus imagination adds up to food anyone can be proud of."[6]

Food had gone from being scarce, expensive, and labor-intensive to becoming a matter of a ten-minute trip by car to a supermarket—cheap and alarmingly conveniently processed and "prepared." For a woman who grew up during the Depression and raised children through World War II, a little convenience made all the sense in the world. Toward the end of her life, however, my grandmother suffered two of the diseases that are linked to our modern diet: type 2 diabetes and cancer.

My grandmother's life was just a microcosm of the changes evident in how we grew, gathered, cooked, and ate our food in just her life span. From when she married in her early twenties to when she died in her eighties, her world transitioned from a regional, seasonal food system that produced 2.3 calories of food energy for every one calorie of fossil fuel it used to a system that took at least ten calories of fossil fuel energy to produce one calorie of highly processed, calorie-rich but nutrient-poor modern food.[7]

We had made an evolutionary leap in our ability to mass-produce food. But at what cost?

A Short History of Industrial Food Production

At the beginning of the twentieth century, the world's population was at 1.6 billion.[8] The general consensus was that the earth had reached its carrying capacity. Voicing a sentiment that was pervasive among industrialized nations, British chemist and physicist Sir William Crookes declared in 1898 that "England and all civilized nations stand in deadly peril of not having enough to eat. As mouths multiply, food resources dwindle. Land is a limited quantity, and the land that will grow wheat is absolutely dependent on difficult and capricious phenomena."[9]

The fear was that there simply wasn't enough food to sustain the .5 percent growth that was adding nine million people per year to the global population. Something had to come along to change the variables, because as they stood, humans had come up against what in today's terms would have been called peak food.

It was already known that nitrogen is a key element in soil that drastically improves plant vigor, growth rate, and crop yield. Farmers would spread compost on their fields because it was rich in nitrogen from the decaying plant matter. They would also plant cycles of legumes like beans, peas, lentils, vetch, and alfalfa, for example, which would also enrich the soil with nitrogen, thanks to a symbiotic soil bacteria that lives on the roots of legumes and makes the nitrogen in dirt available to plants. Lightning striking the ground released the bonds of atmospheric nitrogen, so the rainfall after a lightning strike was a haphazard source of nitrogen, as well. But making compost and growing nitrogen-fixing plants were slow processes, and lightning was far too random.

Saltpeter, or potassium nitrate, another very rich and naturally occurring source of nitrogen, could be added to the soil. But saltpeter was rare, mostly limited to deposits in Chile, which had a near monopoly and could set the price. Moreover, agriculture had to compete for these mineral deposits. The same nitrogen compounds that made terrific fertilizer also made terrific bombs.

The race was on to chemically synthesize nitrogen compounds both for war and for agriculture. Given the population pressures and the political climate at the turn of the century, the stakes were high. Finally, in 1908, the German chemist Fritz Haber figured out how to turn atmospheric nitrogen into ammonia, a compound of nitrogen and hydrogen. He received the Nobel Prize in Chemistry for this discovery in 1918 "for the synthesis of ammonia from its elements."[10] Another German chemist, Carl Bosch, took Haber's technical discovery and figured out how to do it faster and cheaper; that is, he industrialized it by coming up with methods to produce it on a larger scale and keep the unit-cost down. The Haber-Bosch process made it possible to create large amounts of ammonia, and the industrialization push of World War II scaled it up into a very profitable business.

Wayne Roberts, in *The No-Nonsense Guide to World Food*, writes that in the post–World War II era, "the road to junk food, rural poverty and agricultural pollution was paved with good intentions," as Franklin Roosevelt urged the war-weary world to turn its energy toward putting "an end to the beginnings of war."[11] And those beginnings of war were often hunger and food insecurity. Finding a way to provide enough food for all became a global mission of sorts for the United States. And with the war machine's infrastructure already built and operating, it not only became easy to switch over some of the efforts of the military industry to create an agricultural industry, it became an important domestic postwar policy.

With the infrastructure already in place to produce ammonia, munitions factories were tweaked to produce chemical fertilizer, and chemical-warfare science like nerve-gas production was redirected at insects in the form of pesticides. Chemical pesticides, herbicides, and fertilizer became standard practice for commercial agriculture for the first time ever. They were so universally adopted that we now refer to the routine use of chemicals on the farm as "conventional agriculture."

The amount of food that the earth could grow seemed unlimited due to our newfound ability to replace traditional practices like crop rotation and composting, which relied on slower natural rates of renewal and regeneration of the soil's nutrients, with chemical fertilizers. And a population explosion in turn created the need to use more and more chemical compounds to keep up with all those hungry mouths. Economies of scale and regional specialization flourished, with the help of pesticides, herbicides, and heavy equipment, another legacy of modern industrial warfare turned agriculturally inclined.

THE POPULATION BOMB
AND THE GREEN REVOLUTION

Throughout the 1950s, the United States was gripped with Cold War concerns that social and political instability due to famine and hunger in the Third World would provide the conditions for communism. In the name of feeding the hungry and staving off communist expansion, the United States donated billions in food aid to certain nations and exported its industrial agricultural system—high-yield, high-tech grains and the pesticides, herbicides, and chemical fertilizers that supported them—to others. Mexico would be the test case for exporting this Green Revolution, and indeed it changed from a net importer of wheat into a net exporter in the 1950s. The side effect of this Green Revolution, however, was that as larger-scale industrial agriculture replaced traditional subsistence farming in Mexico, millions of independent family farmers could no longer compete in the marketplace. People migrated off the land, into cities, and across the US-Mexico border.

In the 1960s, as the Green Revolution fanned outward, producing record global crops, the global population ballooned with an unprecedented annual growth rate of 2 percent late in that decade. Stanford biologist Paul Erlich ignited a firestorm of panic when his 1968 best-selling book, *The Population Bomb*, predicted apocalyptic scenarios of global starvation in the 1970s and 1980s due to unrestricted population growth, both at home and in developing countries. The book's doomsday scenarios were the push that the US government and its allies needed to rapidly industrialize the agriculture in what was then termed the Third World. Mass starvation due to famine was being predicted for India, and the Indian government invited Norman Borlaug, the American agronomist behind the agricultural transformation in Mexico, to see if he could help avert the worst-case scenario. The Green Revolution succeeded in India, as it had in Mexico.

Human populations have continued to boom hand-in-hand with

industrial agriculture and more-or-less cheap oil. And this has led to a transformed world. We have more than doubled the population of the planet since 1960.[12] In 2008, the global population passed another benchmark. No longer tied to the land, over half of the world's inhabitants were living in cities.[13] In places like North America and Europe, 80 percent of the population is urban. There are now twenty-one "megacities"—urban clusters of over ten million people—around the world.[14] And these megacities are tipping the balance of global power. Writing in *Foreign Policy* magazine, geopolitical economist Parag Khanna notes that "Africa's urbanization rate is approaching China's, and the continent already has as many cities with a population of 1 million or more as Europe does."[15]

In a battle that is still ongoing in places like China, the Indian subcontinent, Africa, Central America, and South America, industrial farming is displacing smaller, more sustainable, adaptable mixed-family farms. Migration to cities is happening at an unprecedented rate. Rural populations are leaving their land with crops selling at such historically low prices that they can no longer afford to stay on the farm. They come to the city fleeing poverty in the countryside, but the life they find in the city is often worse. One in seven people alive today live in slums.[16]

FROM GREEN REVOLUTION TO GENE REVOLUTION

Industrial farming takes from the land faster than even the most liberal applications of high-tech chemical fertilizers can replace. Crop yields are now dropping after a few honeymoon decades as the last of the farmlands' natural capital goes into a final waltz of industrial food production. This is why the industrialized nations are turning to ever-more extreme food-production methods, like genetic modification of crops, even for infinitesimal yield increases.

Since the 1980s, the same companies that brought us chemical fertilizers and pesticides have been tinkering with the crop and livestock at the genetic level to try to bump food production once again. Since 1984, when the first genetically modified plant was produced, 170 different crops have been created in a lab and planted in field trials.[17] And the agribusiness public relations campaigns are an oddly unsettling mix of altruism and apocalyptic predictions of what will happen if we don't accept their genetically modified creations—but it may already be too late to object.

Genetically modified crops have been commercially planted only since 1996, but already, 70 percent of processed foods contain some genetically modified ingredients.[18] The United States currently has 165 million acres (66.8 million hectares) of genetically modified crops planted.[19] Brazil, a relative latecomer to the genetically modified crop game, has 63 million acres (25.4 million hectares) of genetically modified crops.[20] The problem is that many food crops, such as corn, alfalfa, and canola, pollinate when the wind blows or when bees transfer pollen from plant to plant. You can't really control the genetic transfer, so genetically modified (GM) crops can cross-pollinate with non-genetically modified crops if they are in the vicinity. For those who are trying to preserve traditional varieties of certain crops or who want to be certified organic producers, finding genetically modified genes that have infiltrated your crop can be devastating. Farmers in remote regions of Mexico are complaining that they are finding evidence of GM corn in their traditional wild varieties, and they fear losing thousands of years of tradition within just one generation.

Another consequence of GM crops is that "superweeds" have emerged that are now resistant to the chemical herbicides being used routinely in the farming of pesticide-resistant crops like GM cotton, corn, and soy.[21] This mimics the healthcare consequences of the routine overuse of antibiotics, which has created antibiotic-resistant "superbugs."

According to the United Nations, we'll add another two billion people to the face of the planet by 2045. The question on everyone's minds is whether we can produce (and distribute) enough food to feed even more people. Undoubtedly, the schism between the genetically modified, patent-protected corporate vision of the future of food and the open-source, small-scale, broad-based view will get wider as the stakes get higher.

TOO MUCH POWER IN THE HANDS OF TOO FEW

In 2006, the United Nations' Conference on Trade and Development (UNCTAD) published a report called "Tracking the Trend toward Market Concentration: The Case of the Agricultural Input Industry." It identified an alarming narrowing of the players in our global food system. Through "vertical integration," just a few companies own the seeds, fertilizers, herbicides, fungicides, and now even the patented genetic property of the main crops that form the base of the food chain.[22] Bayer Crop Science, Syngenta, and BASF control half of the agricultural chemicals on the global market. Monsanto controls one-fifth of the global proprietary seed production, a 25.2 billion-dollar industry.[23] The report echoed the findings of another report issued two years after the UNCTAD's paper about the alarming concentration of the seed-food distribution business: by 2005, three companies controlled 90 percent of the world's grain trade: Archer Daniels Midland, Cargill, and Bunge.[24]

The large corporate agribusinesses are squeezing the independent farmers to the point of desperation. Yields of genetically modified crops are dropping, but the prices of seed and proprietary fertilizer-herbicide-fungicide combinations stay high. The concentration of agriculture into the hands of a few major multinationals allows them to keep the sell-price of commodity crops low enough that only high-volume producers can turn a profit. Farmers in Pakistan, India, and Korea hope to farm their way out of the debt they have taken on to buy these genetically modified inputs, and many are committing suicide out of despair and to protest a system that they feel no longer has any place for the individual. (Korean farmer Lee Kyung Hae stabbed himself to death in 2003 outside a World Trade Organization meeting in Cancún, Mexico, to bring attention to the victims of globalization—peasant farmers.[25]) As of 2010, the Indian government estimates that there have been over two hundred thousand suicides of farmers in just over thirteen years because of debt and other effects of globalization.[26]

And even if you want no part of this global, industrial, genetically modified food chain, it's virtually impossible to avoid, even if you have your own piece of land and decades of traditional farming experience. Just ask Canadian farmer Percy Schmeiser.

In a decade-long David-and-Goliath court case that went all the way to the Supreme Court of Canada, Schmeiser was found guilty of a type of indirect patent infringement because he unknowingly (and, in his opinion, unintentionally) grew some of Monsanto's patented GM Roundup Ready® canola on his farm in Saskatchewan.[27] In other words, this type of GM canola had been engineered to be resistant to Monsanto's Roundup® herbicide, which could be sprayed on the field and would kill every other plant or weed, except for Roundup Ready canola plants. Monsanto vigorously protects its patent by making farmers who buy the seed agree not to save seeds for replanting, thereby ensuring that the farmers purchase new Monsanto seed year after year, as well as the Roundup herbicide, with which its canola has been engineered to work.

Schmeiser, however, claimed he did not buy seed from Monsanto but rather engaged in the traditional practice of saving some of the seed from the previous year to plant again in the spring. He argued that genetically modified canola from other fields must have contaminated his seed pool over the years. If he had Monsanto's patented canola growing in his field, he never planted any knowingly. He also did not use Roundup herbicide on his canola and therefore didn't benefit from this genetic technology that had crept onto his land.

Five of Schmeiser's farming neighbors grew Roundup Ready

canola. Seed can blow off of trucks during transport or be carried by the wind or even birds. It seems that Monsanto didn't care how its seed got onto Schmeiser's land. The company was determined to make an example of him.

Monsanto took Schmeiser to court in Canada over patent infringement. The ensuing legal battles lasted seven years and went all the way to Canada's Supreme Court, which found that Schmeiser was guilty of growing Roundup Ready canola—albeit unintentionally—without paying the licensing fee. Monsanto used this case to put a chill into the heart of every small farmer who had resisted its aggressive sales tactics, because whether or not you wanted to grow its canola, some might end up in your field anyway. Bending to the bully tactics of Monsanto was really the path of least resistance. These last few stubborn holdouts who didn't buy patented, genetically modified seeds were apparently a thorn in the paw of Monsanto. Outliers, they made clear, would not be tolerated.[28]

Maybe this is another reason that farming is moving out of the countryside and into the cities. To be an independent-minded farmer, you really need to move off the radar of Big Ag. Maybe you need to be too small to matter, and maybe it helps to hide in plain sight—right in the city.

Chapter 3

INDUSTRIAL EATERS

Tell me what you eat, and I will tell you who you are.

—Jean Anthelme Brillat-Savarin,
seventeenth-century French gastronome and
author of *The Physiology of Taste*, 1825

INDUSTRIAL EATERS I: HUMANS

The Fernandezes of Texas have two preteen children: a boy and a girl. Grandma lives with the family. Their weekly groceries and takeout food purchases fan out in front of them on the kitchen table. They include, among other items, breaded fish sticks, extra-lean ground beef, luncheon meat, avocados, tomatoes, bananas, donuts, apples, a head of iceberg lettuce, homemade tortillas, and takeout pizzas. The family's food purchases cost $242.48 that week.[1]

The Revises are from North Carolina. Mom and Dad stand behind the table laden with their week's groceries: bags of potato chips, sodas, a gallon-jug of milk, pork chops, smoked turkey slices, bunches of grapes, and bagged vegetables. The two teenage sons sit with enormous pizza boxes on their laps, and there are other fast-food containers with the rest of the week's worth of food eaten on the go. The Revis family spent $341.98 on food that week.[2]

The Cavens live in California. Dad holds a young boy, no older than four, while his sister cheekily mugs for the camera with her hand under her chin and her head to one side. Mom smiles next to the dad and toddler. Their groceries contain a lot of bread products but also frozen corndogs, frozen peas, and boxes of breakfast cereal. There's also a package of ground beef, some eggs, broccoli, bananas, apples, and tangerines. Their grocery bill for that week is a mere $159.18.[3]

I didn't actually go into the homes of these families and follow them around for a week. They appear, along with their food purchases for the week, in Peter Menzel and Faith D'Aluisio's *Hungry Planet: What the World Eats*. In their award-winning 2005 book, the authors photographed thirty families in twenty-four regions across the world for a pictorial look at household diets. On the pages of the families from countries including the United States, Canada, Germany, Kuwait, the United Kingdom, the United Arab Emirates, Japan, and Mexico, the book captures the diet of industrial eaters.

In industrialized nations, we consume lots of meat and lots of refined carbohydrates, as well as foods that are high in fat, salt, and sugar but low in fiber and nutrients. We tend toward lots of processed dairy yet very few whole foods like fruits, vegetables, seeds, and whole grains. More importantly, we eat too much of everything, usually in a hurry or in front of the television or in the car. Whole grains, fresh fruit, and seasonal vegetables get pushed out of the diet as the cheap, empty calories of processed food take over.

This is what was originally termed the "Western diet" because it used to be associated with the all-American expanding waistline. We now understand that it's a nutritional transition that occurs in most countries as they industrialize. So, what was once considered a uniquely culturally American or Western European diet is now an increasingly Chinese and Indian diet. Given that these two countries make up one-third of the global population, the impact of even small shifts in these countries' eating patterns will be huge shifts in the food ecosystem. Is it any wonder that Walmart is cutting deals to open stores in the notoriously protectionist Indian retail market, as is the UK grocery giant Tesco, and France's Carrefour? Walmart's supercenters in China have already proved to be an extremely lucrative international experiment for the world's largest retailer. The big-box revolution that we experienced in the 1990s in North America and the United Kingdom is going global.

FOOD MILES, FUEL, AND FREEWAYS

The 1990s was the decade when the supermarket took an evolutionary leap in size. It grew hand in hand with increasing global trade and the cheapening of food. Supermarkets became superstores, and big-box grocery stores (even larger than superstores) chased urban sprawl farther and farther out. There seemed to be no limit as to how big our grocery stores could be and how much food from far, far away could be found inside.

Snap peas and garlic from China. Apples from New Zealand. Asparagus and grapes from South America. Strawberries and artichokes from California. Global summertime was in full swing in the grocery store. But it was an apocalypse down on the family farm. Our domestic small- and medium-sized (that is, family) farmers were told that they had to scale up to compete on the global market or get out. Most got out, by way of bankruptcy at the rate of fifty thousand farms per year.[4] Trade liberalization of food crops in the 1990s flooded the global market with ever-cheaper commodity crops, so much so that the cost of seed, fertilizer, and machinery outstripped the price a farmer could get for his crop, even in a good year.

This mass extinction of the last of the family farms in the United States (and in Canada) allowed corporate agriculture to buy up farmland and aggregate immense landholdings. This enabled large operations to maximize the use of economies of scale, which were working only at the commercial farm level anyway. As a result, cheap food became even cheaper. Suddenly we could afford strawberries in January no matter where we lived.

The 1990s was the decade where our food started to travel really long distances. It was the beginning of the global food swap.

The 1,500-Mile Diet

In June 2001, the Leopold Center for Sustainable Agriculture at Iowa State University released a watershed report on the production, transportation, and distribution of food in the US food system. "Food, Fuel, and Freeways" began with the statement that "most consumers do not understand today's highly complex global food system. Much of the food production and processing occurs far away from where they live and buy groceries."[5] Most people, the authors posited, weren't even aware of, let alone concerned about, the drastic increase in the fossil fuel use that the global food system was responsible for. And all but a few were equating Chilean table grapes with climate change and greenhouse gas emissions. The report compiled the research, did the math, and gave the local food movement some significant talking points that are still in use today. It also brought to light the fact that the average grocery store item was traveling absurdly long distances from the field where it was grown to our plates. The center's report also drew the line between these increased "food miles" and the "[e]xternal environmental and community costs related to the production, processing, storage, and transportation" that was seldom accounted for in the food's retail price.[6]

At thirty-seven pages, the report covers an astonishing amount of past research and contemporary concerns about the conventional food system. It was a groundbreaking report in that it took the average grocery store item, something universally understood by every eater in the United States, and laid out the realities of its complex, or at least lengthy, journey from field to plate. It published research on energy use in the food system, with emphasis on the transportation and distribution sector. It compared the distance that food traveled from farm to point of sale in a conventional "integrated retail/wholesale system," that is, the supermarket model, with distances in a local system, that is, local farm co-ops producing for local retailers, as well as farmers who sell directly to consumers via Community Supported Agriculture (CSA) programs and farmers' markets. It used fresh produce to compare miles traveled, fossil fuels used, and carbon dioxide emissions created in the transportation of several food systems. And, finally, it made recommendations for actions to document these "hidden costs" in the conventional food system and provided an argument in favor of local and regional food systems. Perhaps, most astonishingly, it established the most iconic sound bite of our current modern, industrial food item: that the average grocery store item travels over 1,500 miles from farm to consumer.

The fact that this report came out of Iowa is not insignificant. Iowa is an agricultural state, with thirty-three million acres of farmland specializing in the commodity crops of corn, soybeans, hogs, and cattle. By the time "Food, Fuel, and Freeways" was released, these agricultural commodities were almost entirely for export. "With the possible exception of livestock for meat production, most Iowa farms no longer produce food to supply Iowa consumers directly," the report said.[7] Instead, the food produced on the land leaves the state as raw commodities, which then are re-imported into the state once processors have "added value" and, therefore, cost to the food.

There was a time, the authors noted, when Iowa thrived on a regional, diversified agricultural loop. In the 1920s, US Agricultural Census records showed Iowa producing thirty-four different commodities on at least 1 percent of its farms, and ten different commodities on over 50 percent of its farms. Iowa farms produced apples, potatoes, cherries, plums, grapes, raspberries, strawberries, sweet corn, and pears, for example. By 1997, monocrop swathes of corn and soybeans were grown on over 50 percent of Iowa's farms. And with the diminishing diversity of foods, the diversity of local processors vanished. In 1924, Iowa had fifty-eight canning factories in thirty-six counties just to process sweet corn alone. By 1998, there were a total of two canning factories left in Iowa. This once highly agricultural state needed to import 90 percent of its produce.[8]

Food in Iowa, as it does elsewhere, comes and goes largely by heavy trucks burning diesel and spewing carbon dioxide emissions. Between agricultural production, processing, and transportation, the report said that the food system accounted for 16 to 17 percent of total US energy consumption, transportation alone accounting for 11 percent of energy used within the food system.[9] Clearly, reducing the fossil fuels used in our food system would have a significant impact on overall environmental concerns.

The report is the source of the much-used figure of 1,518 miles (usually rounded down to 1,500 miles) that the average item of fresh produce travels from point of production to point of sale on a current calculation of average food miles of food items consumed in the United States.[10] Moreover, the report's authors noted that this type of calculation was something that was getting increasingly more difficult to conduct because government-operated food terminals were vanishing as major food retailers consolidated their operations. Whereas even the chains once bought their produce and fresh meats through these government-owned and operated terminal-food wholesale markets, they have now chosen to create their own distribution systems as they became more concerned with efficiencies and vertical integration. As the food terminals went away, independent grocers lost a vital wholesale supply line. And as the report states, the figure was estimated from production and shipping records for various fresh fruits, vegetables, meats, and other foods that passed through the Chicago food terminal system in 1998. It was much more complicated to calculate the food miles for multi-ingredient processed products.[11] And because so much of the food supply passes through these privately owned systems, it is virtually impossible to get accurate statistics on food miles nowadays.

INDUSTRIAL EATERS II: ANIMALS

We're not the only industrialized eaters in this food chain. We've turned the livestock and the fish we eat into industrial eaters as well. Livestock are no longer a few cows happily munching grass in a field, or pigs wallowing in mud and rooting around in the farmyard. Cows, naturally herbivores, now live in close-quarter pens and are fed soybeans and grain. Swine, whose wild cousins are natural foragers with highly varied diets, are fed monochromatic diets of more soybean and wheat pellets, with a few essential vitamins mixed in. Larger fish no longer chase and eat smaller fish in the aquaculture food chain. Instead, they are fed pellets of fishmeal, cereals, and vegetables.

Concentrated Animal Feedlot Operations (CAFOs) are now the overwhelming norm of meat production in our food system. And they are just as their name suggests: factory farms with a minimum of a thousand head of cattle being fed high-calorie diets to put on weight as quickly as possible in as small a space as possible. Poultry CAFOs have anywhere from ten thousand fryers to one hundred thousand laying hens. Battery hens, also known as laying hens, are kept in wire cages that are so small the hens can't fully turn around. Their beaks are removed by cauterization, because otherwise they'd peck each other to death. Pig CAFOs house upward of 2,500 swine, usually more like 5,000. The largest slaughterhouse in the world is in Tar Heel, North Carolina. Its Smithfield hog-processing plant kills 32,000 hogs a day.[12]

Routine (over)use of antibiotics in CAFOs is part of the reality of industrial meat production because the cramped, stressful conditions that these animals endure leaves their immune systems weakened and prone to disease. In 2009, the first-ever report on the use of antibiotics in the livestock industry was compiled by the US Food and Drug Administration (FDA). In this report, released in 2010, it was revealed that the FDA's best-guess estimate is that 70 percent of the antibiotics used in the United States are in CAFOs for animals, not humans.[13] So while the medical community struggles with new strains of antibiotic-resistant superbugs in hospitals, there's rampant and indiscriminate use of these lifesaving drugs in agriculture simply because it's more cost-effective to grow animals in a space smaller than what allows them to even turn around.

CAFOs, however, are not going away anytime soon. They are the only reason we can afford to eat so much animal protein every single day. And global meat and poultry consumption is set to increase by 25 percent by 2015.[14] The pressure to produce more meat cheaply will come from developing nations. Right now, most of the world's population eats a plant-based diet, but as countries industrialize, so do their diets. This shift to animal protein-rich diets puts even more of a strain on the planet's limited cropland and water resources. Moreover, it diverts a lot of plant-based calories away from the immediate human food chain, into the animal food chain.

To "grow" each pound of meat, the animal must eat ten pounds of feed.[15] In Canada, this means that 73 percent of the grain crop is used for livestock feed.[16] In the United States, it's 60 percent.[17] In the United States, one million acres (400,000 hectares)—an area larger than Germany—are used simply to grow grains to feed livestock.[18] This results in 1.3 million tons of manure, which in CAFOs creates devastating pollution problems. And consider this: it requires 800 gallons (3,000 liters) of water to produce 2.2 pounds (1 kilogram) of dry-weight rice, but it takes a mind-boggling 4,000 gallons (15,000 liters) of water to produce 2.2 pounds (one kilogram) of beef.

We used to produce our beef, poultry, and pork at home, but now we're outsourcing our livestock production to resource-rich yet cash-poor countries to keep the prices down. The South American rain forest—the lungs of our asthmatic planet—is being decimated by one and a half acres per second, in large part to feed our hunger for a meat-based diet.[19] But since it's happening in the developing world, we tend not to have to face or pay for this environmental degradation. Someone else does.

INDUSTRIAL EATERS III: MACHINES

The sudden rise in the cost of food staples during 2007 and 2008, as well as another round of food price spikes in 2011, suggests that we have come to the end of cheap food. But what about that record harvest in 2007, and then again in 2008? Why didn't that at least offset the rising cost of the fossil fuels that industrial agriculture depends on? Shouldn't more supply lower the price in a free market? No, and that's because the food market is now tied to the fuel market.

To guard against a fossil fuel crisis, there was a shift into "green fuels" made from corn, wheat, and other grain crops this past decade. Staple crops suddenly had to compete against fuel prices. Writing in *Der Spiegel*, Lester Brown, president of the Earth Policy Institute in Washington and author of *Plan B 2.0: Rescuing a Planet under Stress and a Civilization in Trouble*, states, "In effect, the price of oil becomes the support price for food commodities. Whenever the food value of a commodity drops below its fuel value, the market will convert it into fuel."[20]

This is especially disturbing, given that in the same article, Brown points out that the grain required to fill a twenty-five-gallon SUV's gas tank with ethanol can feed one person for one year. The grain it takes to fill the tank over the year—assuming two fill-ups a week—could feed twenty-six people.[21] And as we approach peak oil—the point at which the finite oil supplies start their slide down the bell curve of extraction, creating increasing scarcity and inflating value—it will be increasingly profitable to convert our food into fuel. In attempting to solve one problem—fuel shortages—we'll create an even bigger, more serious one.

Even a mere 5 percent of the world's cereals being diverted to agrofuels has caused as much as a 75 percent jump in world grain prices.[22] The whole idea of biofuels has been such a disaster for food prices that the European Union is reevaluating its agrofuels policies.

Paying the Full Price for Our Food Choices

The average American eats about two hundred pounds (ninety kilograms) of meat each year,[23] as does the average Canadian (238 pounds/ 108 kilograms).[24] Every second child born to parents of color after the year 2000 in the United States will get early-onset diabetes; for children born into Caucasian families, the statistics are slightly more favorable: one in three will develop the disease.[25] One-third of the adult population in the United States is overweight, and a further third of adults are obese, according to the US government's Centers for Disease Control and Prevention's most recent statistics.[26]

Yet we pay so very little for all this food we consume. In North America, our food budgets are at an all-time low. Americans, on average per household, typically spend 9.4 percent of their disposable personal income on food purchases.[27] This is the lowest ratio of income-to-food expenditures in the world. But that's because the industrial food system is particularly good at hiding the real costs, like the environmental and social costs that our cheap food is incurring. We're on a "buy now, pay later" program, in terms of both the health consequences of our food choices and the environmental impact.

Raj Patel makes the case in his book *The Value of Nothing* for calculating the real cost for a hamburger.[28] He argues that if a dollar value was assigned to the environmental degradation of industrial meat production, the loss of biodiversity, and the destruction of ecosystems in places like Brazil, that price tag for a burger at the drive-thru would be $200 not $4. (Patel contends that deforestation for soy crops in Brazil, to grow soy bean feed for animals and to create grazing pasture, should be factored into the retail price of a burger.) These are the costs, moreover, that someone else pays, usually the developing world, which is losing its topsoil, having its water tables polluted, and losing its biodiversity to feed overweight Americans with burgers, fries, and soft drinks for sometimes less than five dollars. The cost is real, but deferred, hidden on a lay-away plan, or too often pushed over to be paid by developing countries. It's out of sight and therefore out of mind.

While we may not have to pay the environmental debt directly, we're certainly paying for cheap food with healthcare costs that just keep rising. The obesity epidemic in the United States adds a whopping $147 billion a year to healthcare budgets just for a drive-thru lifestyle.[29] One healthcare dollar of every five in the United States is spent on care for people with diabetes, according to Patel.[30] Yet the US government insists on maintaining obscene subsidies for corn and cereal crops, which enables the very cheap-food diet that has rendered two out of three Americans overweight and strains the healthcare system to the breaking point.

Canada doesn't have such subsidies, but the modern sedentary lifestyle and the fast-food diet is still wreaking havoc. One in four Canadians are obese, taking a $4.3 billion bite out of the $202 billion national healthcare bill.[31]

But since these two figures of healthcare costs and household food costs are calculated in different columns of the ledger, they are rarely part of the same discussion. We're subsidizing agriculture—albeit the wrong parts of it—and then we're stuck with the healthcare bills caused by our overeating. It's comedic, if you can get past the tragedy of it all.

Chapter 4

A WORLD IN FOOD CRISIS

Being an economist can ruin your appetite.
—Jeff Rubin, *Why Your World Is
about to Get a Whole Lot Smaller*, 2010

Strip a system of redundancy, and you increase its efficiency; but you also reduce its adaptability and resilience. Centralize to take advantage of economies of scale, and you create an easy target to disrupt the whole of a system. Strip a food system of diversity and you get the Great Potato Famine.

With profit comes risk. With efficiencies come vulnerabilities. Cheapness for some is paid for by others. With capitalism and trade liberalization, one gets deregulation and speculation. These are the trade-offs that we don't often hear about.

In the era of globalism and free trade, industrial agriculture's promise to produce cheap, abundant food actually came true for a few decades. Our big factory farms produced food at an unprecedented rate, and the costs of farm labor in developing nations meant that food became the cheapest it had ever been. Corn became so cheap that food scientists worked overtime to find new ways to use it. Corn ended up in everything from carbonated beverages (high-fructose corn syrup replaced cane and beet sugar as sweetener) to cosmetics (cornstarch and corn oil) and disposable diapers (cornstarch). Corn also fattens livestock faster than pasture grazing. Steak, chicken, and fish, once luxury items, became daily food choices.

Take Atlantic salmon, for example. Salmon is a sturdy fish that grows fast, even in aquatic farms, and holds up under freeze-thaw cycles. It is well suited to be an industrial product. Economist Jeff Rubin points to the rise in consumption of Atlantic salmon as an example of how cheap oil makes cheap food possible in his best-seller *Why Your World Is about to Get a Whole Lot Smaller*. Salmon caught off the coast of Norway, he writes, are frozen whole onboard a small fishing boat and taken to port. The frozen salmon are then transferred to a larger boat and taken to a larger port, maybe somewhere like Rotterdam or Hamburg. From there they are shipped to China, where the fish are then thawed, skinned, filleted, and deboned by factory workers who make a pittance for a day's work. The salmon are refrozen for the return voyage back west, perhaps ending up in a Costco store in North America or a Tesco store in the United Kingdom. The fact that the fish are affordable, despite two major ocean journeys and months of refrigeration, is a testament to the era of cheap oil and labor we have been living in.[1] But the entire point of Rubin's book is that we are at the end of cheap oil. And as it ends, so does globalization.

It's hard to get worked up about the middle classes of North America losing their weekly salmon dinners when most of the world's population has had the bitter taste of globalization in its mouth for years. Industrial food production was supposed to end world hunger, especially among the nations most at risk of famine and malnutrition, but the number of hungry people has just gone up since the Green Revolution began in the 1960s.

In late 2007 and early 2008, the cost of food rocketed skyward. Despite record global grain harvests in those years, double-digit—even triple-digit—inflation in food prices within weeks ripped through cities in Asia, Africa, and the Middle East. Years of cheap food had actually forced incomes into a downward spiral for those already struggling to get enough to eat. The fuse was the global banking crisis, but it only exposed the underlying catastrophe that had been in the making for years. The United Nation's Food and Agriculture Organization announced that there were 862 million hungry people in the world in 2008 and that the figure would reach one billion in 2009.[2] One-sixth of the global population would be hungry.

In response to this, the Food First Institute for Food and Development Policy in Oakland, California, issued a policy brief in October 2008 titled "The World Food Crisis." It pointed the finger at the "decades of skewed agricultural policies, inequitable trade, and unsustainable development" that has finally "thrown the world's food systems into a volatile boom and bust cycle," where the gap is now widening between the rich and the poor.[3] The paper's author and Food First Institute's executive director, Eric Holt-Giménez, also challenged the myths that are behind the huge lobbying efforts of corporate agriculture to continue the legacy of the Green Revolution into the genetically modified foods revolution—namely, that the world must employ industrial agriculture and genetically engineered crops and livestock to feed the growing population. (*Of course* corporate agriculture and biotechnology businesses are going to insist that the world cannot support its growing population without their proprietary technologies.) However, Holt-Giménez points out that production is not the problem, cost and distribution are. Global food production has actually increased by 2 percent a year over the past twenty years, while population growth has slowed to 1.14 percent. "Population is not outstripping food supply. People are just too poor to buy the food that is available."[4] As food has become cheaper, people's purchasing power, especially in developing nations, has diminished even faster. Three billion people worldwide live on less than two dollars per day.[5] *Cheap* and *abundant* is irrelevant to people who can only afford to window-shop.

More than thirty countries experienced food riots in late 2007 and 2008. These riots often turned violent. And just because there haven't been food riots in the United States doesn't mean that the world food crisis hasn't been taking its toll on Americans. Food banks were being overwhelmed. In the United States, 50.2 million people—15 percent of the population, including 17.2 million children—weren't getting enough to eat.[6]

This was just the first wave of the global food crisis, as it turned out. The second wave was touched off by weather events in just a few key food-growing regions. Russia, which accounts for 11 percent of global wheat exports, had a heat wave on top of a drought in the summer of 2010.[7] Record high temperatures and wildfires devastated its wheat crop, and President Vladimir Putin announced a nationwide ban on wheat exports. Australia, the world's fourth-largest exporter of wheat and a source of another 11 percent of global wheat exports, experienced flooding over an area the size of France and Germany combined in January 2011.[8] Australian sugar exports had to be reduced by 25 percent because of the flood damage as well.[9] The United Nations reported that global food prices had risen above the 2008 record, and, predictably, deadly riots began again in early 2011.

We have arrived at a point at which industrial agriculture can no longer continue to make good on its promise of cheap, limitless food. The variables have changed, and the system is breaking down. We are losing cropland and predictable weather cycles through climate change. We're using up our fresh water and land at unprecedented rates. And we're coming to the end of cheap oil. We have reached the end of the era of cheap food.

CLIMATE CHANGE

The earth's temperature has been warming up significantly since the 1950s.[10] This is primarily a result of our industrious, transportation-crazy species. We've been burning fossil fuels that emit carbon dioxide into the air, and we've been chopping down and burning our forests to make way for more agricultural and residential land. (In a balanced ecosystem, kingdom *Plantae* turns carbon dioxide into oxygen for kingdom *Animalia*, which takes in oxygen and breathes out carbon dioxide in a nice, tidy cycle. Cutting down the great global forests and building carbon dioxide–emitting machines is definitely upsetting the natural balance.) Moreover, we've been piling our garbage in giant heaps that produce methane as they eventually decompose. Methane, like carbon dioxide, is a greenhouse gas. Together, they are the two main culprits for the steady global warming over the past several decades.

Climate change is creating havoc in traditionally dependable food-growing regions with significant weather events. It's not just the rise in average temperature; it's also the violent natural events that seem to be on the rise with increasing climate change. Not only are droughts, floods, freezes, fires, and changing weather patterns taking their toll on the food supply; global warming as a whole is threatening to shrink our ability to grow food. In "The Future of Food Riots," a January 2011 article syndicated in newspapers around the world, Gwynne Dyer, author of the 2008 book *Climate Wars*, wrote, "The rule of thumb is that we lose about ten percent of world food production for every rise of one degree Celsius in average global temperature."[11] In an even more dire prediction, Dyer continues, "So the shortages will grow and the price of food will rise inexorably over the years. The riots will return again and again."[12]

With such high concentration of key crops growing in specialized regions—wheat in Russia and Australia, tomatoes in California and Florida, citrus in Brazil—crop failures in just one location can cause a worldwide shortage and steep price increases even with a relatively localized disaster.

Historically, nations took their food security into their own hands by keeping food stores in reserve. Grains have traditionally been kept in storage to serve as famine insurance. Yet national grain reserves were seen as unnecessary, even anti–free trade, in the globalization frenzy of the 1990s. Most governments bought the line that the global market would provide, so stockpiling national grain reserves was unnecessary in the new global economy. Developing countries were pressured into privatizing or selling off their national grain reserves in order to qualify for debt relief and other international funding by the World Bank and the International Monetary Fund. With agricultural subsidies flowing freely from the United States and European governments, the cost of commodity crops on the world market was so low that farmers in nonsubsidized nations, such as those in Central and South America and Africa, couldn't compete. This allowed the big grain companies in the industrialized nations to take control of the global grain trade.

Now, as the world was entering this current food crisis, Holt-Giménez calculated annual profits of some of these companies and found that Archer Daniels Midland's profits were up 38 percent from one year to the next, Cargill's were up 128 percent, and Mosaic Fertilizer, a Cargill subsidiary, posted profits of 1,615 percent from 2006 to 2008, just as the economic crisis was coming to light.[13]

Now, anytime the world's supply of grain dips below seventy days, which it is often doing, the cost goes up dramatically, making the big multinational corporations that control grain supplies, seed, fertilizer, herbicides, and pesticides obscene profits. Scarcity is very good for business if you are a capitalist. And there's even a term for this: disaster capitalism, because companies have learned that disasters can be highly profitable.[14]

Peak Land

It is thought that 90 percent of the earth's arable land is already in use. Cutting down forests is one way to create more farmland; buying up cheap virgin land or underutilized (which sometimes actually means sustainably farmed) farmland in Africa, Central and South America, parts of Asia, and in crisis-bound countries is another.

The Daewoo Logistics Corporation, a major shipping and transportation company in South Korea, envisioned itself a natural resource development player and began investing significant energy and money into Madagascar. By 2008, it was working out a deal with Madagascar's government granting Daewoo a ninety-nine-year lease for 3.2 million acres (1.3 million hectares) for produce corn and palm oil plantations.[15] It was going to be the biggest deal of one single foreign country securing access to farmland in Africa since the current food crisis began.[16] And it would provide South Korea with a secure source for half of its own corn needs, which it was getting at the time from the United States and South America. The deal was scuttled when it became known that it would have granted Daewoo access to half of the farmland in Madagascar. Massive public protests forced the government's hand, and the deal was cancelled. Daewoo Logistics Corporation had obviously invested heavily and got burned. The company was banking on huge profits with a secure source of corn and palm oil. It went into receivership in July 2009.

Perhaps it was the audacious scale of this land grab that was its downfall. Yet other land deals are happening today at an unprecedented rate, especially in Africa. Very few governments seem to be paying attention.

Every day, thirty tons of tomatoes and hundreds of tons of pristine, fresh produce leave Awassa in Ethiopia for Addis Ababa.[17] It is the country's largest greenhouse operation at 50 acres (20 hectares), and it sits on 2,500 acres (1,000 hectares) of land leased for ninety-nine years. The country's capital is by no means the produce's final destination. None of this produce stays in Ethiopia. Instead, the very same day, or the next day, it's in shops in the oil-rich, water-poor countries of the Middle East. The greenhouse is owned by Saudi billionaire Sheikh Mohammed Al Amoudi. He is just one of the "petromonarchs" shifting his investments from oil fields to tomatoes and potatoes.

This is apparently the first phase of a plan to acquire 1.2 million acres (500,000 hectares) of land in Ethiopia over the next few years to grow rice, wheat, vegetables, and flowers for Saudi consumers. Commercial farms are popping up in Ethiopia because the government is willing to lease 7.5 million acres (three million hectares) of its most fertile land at a reported price of one dollar per 2.5 acres (1 hectare) per year. Ethiopia is also a country synonymous with famine and food aid. There are still thirteen million citizens who are in need of more food than they currently are getting.

Ethiopia is a particularly jarring example, but it's not the only one. Over 125 million acres (50,500 hectares) worldwide are being "outsource farmed" by rich countries. Japan and South Korea import 70 percent of their grains, and Africa is the site of a neocolonial land grab. It has fertile farmland, water resources, and desperate or corrupt governments ready to sell farming rights to the highest bidders. (Because Africa was spared the industrialized farming revolution that transformed the great farm belts of India, China, Mexico, and the United States, many countries in Africa are now seen as the last frontier of viable, nutrient-rich farmland. And it's going for cheap. Land in Africa is leasing for one-tenth the price of land in Asia. It is no surprise that wealthy countries are on a major shopping spree for agricultural land leases in places like the Sudan, Nigeria, Tanzania, Mali, Uganda, Sierra Leone, Ghana, Kenya, and Malawi, to name a few.

In keeping with the arrival of the large commercial foreign farms, subsistence farmers and small family farms are having their land and livelihood sold out from under them. Those who do get paid are getting the equivalent of ten years of crop yields. This may sound like a healthy sum for poor farmers, but really it's just the equivalent of a half-generation's income. And this is being sold to African nations as "tools for development," to potential investors as the new strategic financial asset, and to the rest of the world who cares to pay attention as the only way to feed the projected nine-billion-person global population by 2045 through high-yield genetically engineered organisms, monocrops, and commercial farming. It's enough to make you believe in mass delusion.

It's not just food security that is driving this land grab. One international nongovernmental organization (NGO), called GRAIN, reports that the world financial and food crisis of the past few years has caused "all kinds of actors from the financial and the agribusiness sectors—pensions, hedge funds, etc." to abandon the derivative markets, "and the food and financial crises combined have turned agricultural land into the new strategic asset."[18] Pension funds are even investing in farmland.

These extraterritorial agricultural annexes are not just in Africa; they are happening wherever anyone can place a "for sale" sign on some cheap farmland, like in the American Midwest and on the Canadian Prairies. The ongoing farm crisis at home is no longer front-page news, but it should be. American and Canadian farmers continue to post record losses, surpassing those of the Great Depression. And in the wake of foreclosures and crippling debt, strategic asset companies specializing in accumulating farmland are nipping at their heels. These companies are snapping up family farms at fire-sale rates to increase the profits and ownership of international speculators. The sentimentally named Walton International Group owns 60,000 acres (24,000 hectares) of farmland in the wheat belt of Canada and the United States.[19] The investment capital comes from Germany, Japan, and the Middle East. So, what may look like medium-sized or family farms bucking the trend and hanging onto their livelihoods by their fingernails may already be tenant farmers. How can it be otherwise when the average American chicken farmer will have to invest over $500,000 in poultry farming start-up costs yet will make only $18,000 per year?[20] Or when wheat routinely sells at a dollar less per bushel than the cost of production? If what we are after is cheaper and cheaper food, this means that we are purposefully driving our family farmer to the poorhouse and eventually out of existence.

PEAK OIL

"Peak oil" does not refer to the time at which the very last drop of black gold is scrounged from the last oil well on the planet. Rather, it's the point at which the resource extraction reaches its maximum capability. It's the top of the bell curve of global production, not a cliff that rises and then plummets vertically straight down from its peak. After cresting at record high levels of supply and record low levels of cost, each barrel thereafter gets exponentially harder to tap, and oil production enters into a state of terminal decline. Eventually decline becomes depletion, and oil becomes really expensive as those last dribbles are extracted.

Peak oil is not a universally accepted idea. Some believe that where there is a demand for oil, someone will always find a supply. This certainly was the case when companies started to dig up vast strips of northern Alberta to extract oil suspended in tarry sands, a proposition only viable when the cost of a barrel of oil stays above sixty-five dollars.

But most economists, geologists, and governments do believe in peak oil, and there is no disputing that the industrialized world runs on fossil fuel. Everything from heating our homes to growing, processing, transporting, and selling the food we eat is dependent on it. But oil is a finite, and now dwindling, resource; we're already casting about to find new efficiencies or even a new source of energy to replace our oil diet. Some peak oil pundits are convinced we've already crested; others predict it will happen by 2020 unless there are significant discoveries of new oil fields.

There's no end to the speculation about the end of oil and the mayhem that will result when there really is no more oil left in the ground. University of Guelph professors Evan Fraser and Andrew Rimas describe a rather grim scenario in their *Empires of Food: Feast, Famine, and the Rise and Fall of Civilizations*: "Should the derricks stop pumping crude, the loss of fertilizer would drop crop yields by half. Three billion people would lose their daily sustenance."[21]

It should be said that many brilliant minds are cautious about doomsday end-of-oil scenarios. Other forms of energy, they feel, will step up and take oil's place, whether it's wind energy, solar energy, or other geological resources.

Vaclav Smil is a distinguished professor in the Faculty of the Environment at the University of Manitoba and an international critical thinker with interdisciplinary expertise in energy, environmental, food, population, economic, historical, and public policy studies. In *Energy Myths and Realities*, Smil lays out the case that the process of switching from fossil fuels to another source of energy will not take place over a few years but rather will come at a significant cost and will occur over a much longer timeline than most people are aware of. Smil argues that comprehensive energy transitions take several generations.[22] For an overwhelmingly oil-dependent food system, as we have now, a several-generation transition is a dire prediction indeed. And our early attempts to find a suitable replacement—such as biofuels—have not gone so well.

PEAK WATER

Growing food demands a tremendous amount of fresh water. The figures are almost unbelievable. The daily food needs for one person can require up to 1,320 US gallons, or 5,000 liters.[23] Of all the fresh water used in the world right now, 70 percent is used by agriculture.[24] And the United Nations has identified that water, specifically the scarcity of fresh water, will be the defining resource of the next century. We currently go to war over oil reserves; it's not unlikely that we'll go to war over fresh water access in a few generations.

The United States depends on fresh water flowing to it from Canada, just as Egypt is dependent on water flowing in from its neighbors. Shutting off the tap from the water-rich countries to the water-poor countries would result in catastrophe. And moving this water around to where it is needed for agriculture requires energy. One-fifth of the energy used in California is simply to get water from one place to another. Water resources, therefore, are also tied to energy resources.

China prefers to use its abundant river water in the north for more profitable industrial purposes than for growing food. The trade-off is that it must rely on wheat imports and the fluctuations of the global wheat market.

Middle Eastern countries are outsourcing the farming of water-intensive staple crops like wheat to Africa. The Saudi government has a stated goal of reducing its domestic grain and cereal production by 12 percent simply as a means of conserving fresh water.[25] Countries that are exporting food are not only exporting wheat, beef, apples, and heads of lettuce; they are exporting their water resources. When water becomes scarce, it will only add to the cost of food, and water-rich countries will be holding the cards.

PEAK FARMING KNOWLEDGE

One issue that gets very little attention in this whole discussion around industrial agriculture is that the world is losing farming knowledge faster than it is losing natural resources. There are currently about 2.8 billion farmers left on our planet of 6.7 billion inhabitants.[26] Now one industrial farmer feeds over 140 people on average.[27] Farmers account for a mere 2 percent of the US population.

Most of the planet's farmers are peasant farmers and subsistence farmers. Many are quite poor. Many live in places where land registries are nonexistent. The security of knowing they will be able to continue to farm their land is tenuous at best. When they are displaced, the knowledge of the specific food-growing skills in their regions will be lost within a generation. It will be a loss of traditional knowledge that has taken tens of thousands of years to accumulate.

THE PERFECT STORM

With all these elements in play, it's actually a wonder that the global food crisis caught most of us off-guard. Already by 2006, investment banks were turning to commodity markets—especially crops—rather than derivatives and other speculative financial market options. Wheat and other grains were very much in demand as biofuels. The word *agflation* became the term the international financial markets used to explain the link between the rise in general inflation as the cost of commodity crops went up, which until this point were seen as external factors in general inflation. Asia and Africa felt the brunt of these increases. When the price of wheat doubled and other staples soared with inflation, the landlocked West African country of Burkina Faso erupted in riots in three of its major cities in February 2008. Soon there were more food riots in Africa and the Middle East, notably in Egypt and Yemen.

Then food riots began bubbling up in Europe, the Americas, and Asia. The rise in price of wheat sparked "pasta riots" in Italy. And the inflation in the corn markets caused "tortilla riots." Argentina had a "tomato boycott" to protest the soaring costs of its favorite vegetable. Brazil's government also announced that it would stop exporting rice for the time being. When the price of rice doubled in two weeks in March 2008, food riots broke out in Asia as well. In 2009, the cost of potatoes in India went up 136 percent, and other foods like chickpeas, lentils, and beans went up over 40 percent.[28]

The next time the price of food rose so sharply, in 2011, the violence increased and governments were brought to their knees. Egyptians took to the streets over the increase of the price of wheat, and therefore bread, a daily staple. It's still underreported in the media that food shortages and rising costs of staple foods like bread were the spark for so much of the political unrest that is unseating governments in the Middle East.[29]

In 2011, the United Nation's Food and Agriculture Organization released another devastating Food Price Index indicating that food prices globally rose above the record peak of 2008. A second wave of widespread food riots began in cities in Tunisia, and the long-time ruler of Tunisia was chased out of office. (It's no surprise that the food riots took place in the cities. Urbanites are always more vulnerable when there are food shortages or price hikes than are their rural counterparts.) Then the unrest spread to Bahrain, Algeria, Jordan, Yemen, Albania, Syria, Lebanon, and then Egypt. Soon it moved on to Libya, the Sudan, and the Ivory Coast. Hunger is a surefire igniter of change in even the most despotically ruled places.

So it seems that we are back to where we began at the start of the twentieth century, convinced that the earth has reached its carrying capacity and looking for the next agricultural leap that will feed the billions more mouths predicted to be born in the next half century. Big Ag is telling us that we need to rely even more on their patented "life sciences" technology—expensive, proprietary genetically engineered seeds and livestock, chemical pesticides, herbicides, fungicides, antibiotics, and fertilizers—even though this new technology is homogenizing entire rural landscapes in the service of producing a very narrow range of foods. But there is another solution to the world food crisis already germinating. It looks nothing like an endless genetically modified soybean field that stretches to the horizon where a rainforest stood only years ago. In fact, it's everything that Big Ag is not. It's decentralized, open-source, small-scale, incredibly diverse, scattered, and even a little chaotic. It's able to produce without heavy equipment or expensive infrastructure. And it's within arms' reach. It's food being grown, distributed, shared, and eaten right in the city.

Chapter 5

THE NEW FOOD MOVEMENT AND THE RISE OF URBAN AGRICULTURE

We are all co-producers.
—Carlo Petrini, founder of the Slow Food
movement (statement made in 2007)

THE FIRST WAVE

In November 1992, a lanky, soft-spoken academic type appeared on Britain's Channel 4 television magazine called *Food File*. Tim Lang had just a few minutes to sound off in a segment called "Mouthpiece" on a topic of his choosing. As the chair of the Sustainable Agriculture, Food, and Environment (SAFE) Alliance, he chose to bring attention to the realities of the globalized food system, something that very few people were even aware of at the time. A lot of their food, not just the obvious exotic fruit they bought at a supermarket, came from other countries, not to mention other continents.[1]

Lang knew about the intricacies of the food system and had serious concerns about where the increasingly globalized food chain was heading. He had been a small-scale livestock farmer in the 1970s before moving into the academic realm in the early 1980s. From 1984 to 1990, Lang was director of the London Food Commission, one of the early municipally created and funded food-policy councils.[2]

In the 1990s, Lang campaigned relentlessly to highlight the absurdity of the global food trade, hoping to bring about awareness and change. He first urged policy makers, especially those steering the General Agreement on Tariffs and Trade (GATT), the forerunner of the World Trade Organization (WTO), to consider the hidden and unforeseen ecological, social, and economic consequences of trade liberalization. The quest for lower and lower food prices was resulting in complex routes and absurdly long distances that common grocery store items traveled between production and the store purchase. No one, it seemed, was weighing the potential environmental, health, and cultural costs. Instead, policy makers were marching forward with the idea that open markets equaled progress.

Lang wondered if he could influence the discussion by targeting consumers, making them aware of what was being left off the labeling. Lang and colleagues at the SAFE Alliance were already talking about "food miles" as shorthand for the distances that foods were traveling between the point of production and the end consumer. Perhaps it was just the term to "help consumers engage with an important aspect of the struggle over the future of food—where their food comes from and how."[3]

Lang felt strongly that consumers needed to become aware that there had been an out-of-sight revolution in how (and where) food is grown, processed, distributed, and sold, and that they were putting this revolution into their mouths every single day. This short television segment was his opportunity to take what had to this point been an insider's knowledge of the global food system and plant the seed that people should consider their grocery choices not only by price or appearance but also by their food miles. Handily, the term was perfectly suited to a consumer-friendly sound bite on television.

In the televised segment, Lang and cameras visited Covent Garden Market, an enormous food hall for fruit, vegetables, and flowers in central London. Food vendors at their stalls were interviewed, as was the head of a major national grocery chain. The finale featured Lang, filmed from fifty feet (fifteen meters) above, pushing a shopping cart around a huge map of the world, choosing items from around the globe, and accumulating food miles with each purchase. Lang then proceeded to do a similar shopping trip of mostly British and European foods, tallying up a much lower food-mile total in contrast.

"I was able to suggest to people watching that they might like to judge their food, not just by price or what it looked like, but also by its food miles, how far it had travelled," Lang wrote a decade later in the May 2006 issue of *Slow Food*, the journal of the international Slow Food movement. "It looked surreal and was intended to be amusing. The programme ended with me suggesting a tax on food miles to curb absurd and damaging energy use."[4]

Lang soon realized that his Benny Hill–esque moment on national television hit a nerve with British shoppers. By 1993, the term "food miles" had entered the mainstream lexicon and was used in major newspapers in the United Kingdom.[5]

In 1994, the SAFE Alliance fanned the flames of food-miles fever in the United Kingdom with fleshed-out and concretized research headed up by Angela Paxton in *The Food Miles Report*.[6] And the Sustain Alliance, which was born from the merger of the SAFE and National Food Alliances in 1999, followed up with another groundbreaking report, *Eating Oil*, in 2001.

Meanwhile in Italy, another reaction to globalization was taking shape. In 1986, McDonald's opened an outlet beside the Spanish Steps, one of Rome's major landmarks and public gathering places. This was the last straw for many locals who saw their great food culture being invaded by the Big Mac®, of all things, creeping in from America. As Carl Honoré deftly puts it in his 2004 book, *In Praise of Slow: How a Worldwide Movement Is Challenging the Cult of Speed*, this was "one restaurant too far: the barbarians were inside the gates and something had to be done."[7]

Carlo Petrini, a visionary food journalist, saw the future, where the flavors and customs of Italian life would be overtaken by the cheap, the fast, and the predictable. Petrini and fellow fast-food opposers chose the name Slow Food for obvious reasons as their visceral reaction to this affront of globalization and homogenization of food culture. By 1989, he'd gathered equally outraged gastronomes from fifteen countries to the founding Slow Food meeting in Paris, where they rallied around, endorsed, and adopted the Slow Food manifesto written by Italian poet Folco Portinari. It began:

> Our century, which began and has developed under the insignia of industrial civilization, first invented the machine and then took it as its life model.
>
> We are enslaved by speed and have all succumbed to the same insidious virus: Fast Life, which disrupts our habits, pervades the privacy of our homes and forces us to eat Fast Foods.
>
> To be worthy of the name, Homo Sapiens should rid himself of speed before it reduces him to a species in danger of extinction.
>
> A firm defence of quiet material pleasure is the only way to oppose the universal folly of Fast Life.

But concern about food miles, awareness of disappearing traditions around food, and lyrical opposition stances were one thing. What were people actually *doing* about the situation?

THE SECOND WAVE

This first wave awakened pockets of enlightened eaters. As we moved into the 2000s, consumers began to vocalize their concerns over industrial and long-distance food chains. Shoppers in the United Kingdom started demanding that major grocery chains carry more local and regional foods. Food-mile fever in the United Kingdom trickled up from consumer to the heads of the big grocery chains, in a reversal of the top-down decision making about people's food choices that had been dominating until then. It was an example of how a broad-based consumer movement could effect real change that the grocery industry had to respond to.

Slow Food's philosophies and adaptable organizational structure traveled surprisingly well outside of Italy. Eating locally, ironically enough, became a global movement, and Slow Food ideals and *convivia*, as its chapters are called, took root early on in places like San Francisco, Portland, Seattle, and New York.

Farmers' markets, which had nearly gone extinct from the 1970s to 1990s in the United States and Canada, came roaring back with surprising vigor. By 1998, there were 2,756 farmers' markets operating in the United States.[8] By 2009, the number was up to 5,274, clearly a weekend routine for many urbanites by now.[9]

Furthermore, Michael Pollan wrote the right book at the right time. *The Omnivore's Dilemma*, which appeared in hardcover in 2006, exposed the realities of the industrial food landscape and how so much of supermarket stock is arguably not so much food as it is food science, and it laid bare the effects the industrial food culture we live in has on our bodies, our families, and our society. All of the sudden, I had friends, who never before cared about what went in their mouths, heading out to farmers' markets and asking me where they could get local, organic chicken, beef, and pork.

By purchasing food directly from farmers, Community Supported Agriculture (CSA) programs—where households pay money up front to a local (usually small-scale) farming operation in return for a share of the produce from early spring to late fall—were beginning to emerge alongside farmers' markets as another alternative that was "beyond the barcode." The United States Department of Agriculture (USDA) reports that there were 400 CSAs operating nationally in 2001; by the end of the decade, there were more than 1,400.[10]

The Slow Food agenda progressed from education to activism, as its influence reached around the globe. It turned its efforts toward biodiversity preservation, promoting fair wages for agricultural endeavors, supporting family farmers, and preserving traditional food knowledge. Rather than shrinking from the root causes of the problem, the new Slow Food philosophy squarely opposed "the standardization of taste and culture, and the unrestrained power of food industry multinationals and industrial agriculture."[11] Now, with members in several dozen nations, it streamlined its message to the more politically charged "good, clean, and fair food for all."[12]

By the end of the 2000s, Carlo Petrini famously urged Slow Food members around the world to cast off their passive title of "consumer." In its place, he deputized them "co-producers," effectively empowering them to exercise their political and economic role in the changes they wanted to see without sacrificing Slow Food's pleasure-focused rationale. "What's good for you is good for the world!" Slow Food promises on its website.[13] Eat better, spend money on food, open yourselves up to pleasure, and save the world at the same time.

General consumer awareness had definitely come a long way in a few short years. There was no denying that discussions about where our food was coming from had gone mainstream. At the end of 2007, the *New Oxford American Dictionary* proclaimed *locavore* its word of the year.[14] Locavores are those who buy their food from farmers' markets rather than from supermarkets for both environmental and food-quality reasons. Local eating, at least in name, had entered the mainstream.

THE THIRD WAVE

Every generation, I suppose, feels that it is living in changing times. But statistically, on a planetary scale, we *are* living in a world that has shifted in the most fundamental of ways.

In 2008, the United Nations Population Fund reported that humanity had passed "an invisible but momentous milestone."[15] For the first time ever, more people on the planet were now living in cities than in the countryside. By 2030, over two-thirds of our planet's population will live in cities. Already 80 percent of the populations in industrialized countries such as the United States, Canada, and the United Kingdom were living in urban settings.[16] We are now an urban species, living in an urban habitat, and the way in which we choose to organize our basic necessities of food and shelter will be affected accordingly.

While we may have broken ties with rural living, we haven't broken our ties with our need to eat. If the second wave, as personified by Michael Pollan's journey through the various food chains in *The Omnivore's Dilemma*, asked "What are we going to have for dinner?" then the third wave is asking, "Where will this dinner come from?"

For the longest time, we've planned our cities around transportation needs, housing needs, recreational needs, and sanitation needs, all the while hoping that the rural lands around us would continue to produce food and that the cheap fuels would continue to flow to transport it from farther and farther away. Outsourcing our agriculture to other continents is looking like the last gasps of fuel-intensive industrial agriculture. Energy conservation will drive us to shorten that global food chain. It all leads us back to the city. We are just starting to rethink our cities *deliberately* with our food needs in mind. (For an excellent take on the intrinsic relationship between food and cities, read Carolyn Steel's *Hungry City: How Food Shapes Our Lives*.[17]) Cities have resources like land, water, labor, and a ready-made market for food production. It actually makes a lot of sense to shorten our food chain by growing food right in the cities where we "co-producers" live.[18] And we are just waking up to the possibilities of what might happen if we consciously plan our cities with food in mind.

Traditionally, agriculture and food have been the purview of larger regions, like states and provinces, as well as national governments. But we are entering the era of the megacities, where clusters of tens of millions of people are living in continuous metropolitan landscapes. It's not surprising that cities, not nations or national agendas, are now the new drivers of change in the twenty-first century.

Urban agriculture, which this book is intently focused on, is going to change how cities are planned; how they work; how they look, smell, and feel. And with our rapidly urbanizing population, how cities feed themselves is going to be the defining obsession of the current century. And urban food production and local distribution chains are now emerging as major game-changing factors in how we city dwellers live and eat.

There are currently some eight hundred million people around the world engaged in some sort of urban food production, growing an estimated $500 million worth of fruits and vegetables—figures that make it seem as though urban agriculture is taking the world by storm.[19] Yet these alternative food systems of urban agriculture, CSAs, and farmers' markets comprise a fraction of 1 percent of the food landscape in industrialized nations. The USDA reports that less than 1 percent of food sales in the United States are direct from farmer to consumer, via farmers' markets or CSAs.[20] There really are no numbers available for the fraction of food retail resulting from urban agriculture at this early stage.

The remaining percentage of food sales is still coming through the supermarkets and the supply chains, where the primary concern is cutting costs, and so those lowest prices are sourced from the global commodity markets and highly specialized industrial food chains.

In other words, we have a long, long way to go to revamp our food systems, but we *are* starting to rethink how we will feed our mostly urban population in a post–peak oil, post–peak water, and even post–peak land future. As the prolific author and food-justice advocate Vandana Shiva points out in *Soil Not Oil: Environmental Justice in an Age of Climate Crisis*, "No society can become a post-food society."[21] Or, as even the most despondent farmer will tell you, we'll always need to eat.

Chapter 6

PARIS

The Roots of Modern Urban Agriculture

Il faut cultiver notre jardin.
[We must tend our garden.]
—Voltaire, *Candide, ou l'Optimisme*, 1759

Perhaps my favorite moment from all my travels for this book happened while I was walking down a tree-lined Parisian boulevard. A jet-black, boxy police station, complete with black-tinted windows, stood out from the jumble of new and old architecture around it. The station was an excellent example of 1970s Brutalist architecture, designed to intimidate and terrorize. Someone, however, had planted grapevines in the window wells around the first level off the sidewalk, and the saplings were gamely doing their best to climb this slippery monolith. I burst out laughing at the juxtaposition of these vines struggling to soften the intentional harshness of this impersonal, aggression-inspired building. Neither the vines' efforts nor the efforts of the gardener who planted them were in vain. The vines were winning.

THE BIRTH OF FRENCH INTENSIVE AGRICULTURE IN NINETEENTH-CENTURY PARIS

French urban agriculture evolved from the walled gardens of medieval Paris to its apogee in nineteenth-century Paris. High stone walls were built to enclose gardens, protecting valuable produce and absorbing heat during the day in order to release it back into the garden areas at night. These sheltered enclosures were microclimates for delicate food plants, not only extending the growing season but also allowing for early ripening of fruit and almost year-round growing of even the most tender vegetable crops. The soil was continually built up in raised beds (both for heat retention and drainage) through liberal dressings of the horse manure that Paris had in abundance, making use of what otherwise was a major transportation pollution problem at the time. Not only were nutrients returned to the soil, but the heat produced as the manure composted down created fermenting hotbeds, keeping the frost off delicate plants and extending the growing season to almost year-round. The use of heat retention with walls and creating heat through the manure used in the raised beds were two key reasons so much food could be grown out of season, fetching much more money for the market gardeners.

Glass-topped "cold frames" placed over crops, or the bell-shaped glass covers (known as *cloches* in French) were also widely used to force early growth and ripening of high-value food crops. Dense plantings reduced watering requirements, and companion planting—that is, using natural symbiotic relationships in plants to enhance growth and food production—was developed to keep both weeds and pests under control. (Certain plants and flowers repel common pests; for example, chives, onions, or even marigolds repel aphids; sunflowers attract aphids but can be used to keep them off tomatoes and artichokes. Some companion plants are used as structural support: corn can provide a "pole" for pole beans, and squash is a common low-growing groundcover that will keep weeds down in corn and other tall plants.)

This combination of efficient growing techniques, which later became known as French Intensive Agriculture, is still in use, though it is often called "square-foot gardening" or *potager* gardening. And it is what allowed the city's fresh fruit and vegetable production to flourish even as the city's population doubled from one million to two million in the last half of the 1800s.

Le Marais is a historic district of Paris on the right bank of the Seine River—now also known as the Third and Fourth Arrondissements, as the municipal subdivisions of Paris are called—and was the epicenter of the city's urban-agriculture industry. (*Marais*, French for "swamp," is a holdover from when this area was boggy marsh. Though the swamps were drained in the twelfth and thirteenth centuries, the name stuck.) Its legacy is that market gardeners throughout France are known as *maraîchers* or *maraîchères*: essentially, swamp gardeners.

Not a lot of attention has been paid to these origins of intensive urban agricultural methods now employed from Cuba to Canada. The best account appeared in an academic agricultural journal, *Agro-Ecosystems*, in 1977. Dr. Gerald Stanhill's paper "An Urban Agro-Ecosystem: The Example of Nineteenth-Century Paris" calls it "one of the outstanding urban agro-ecosystems—the 'marais' of Paris, during the second half of the nineteenth century—the period of its maximum importance."[1]

By the second half of the nineteenth century, Stanhill reports that an estimated 8,500 urban farmers, or *maraîchers*, were working just under 3,500 acres (1,400 hectares) in the city—one-sixth the area of Paris at the time. These small urban farms supplied the city year-round with one hundred thousand tons of high-quality, high-value salad and vegetable crops, with enough surplus to export to England. Each 2.5-acre plot (1 hectare) was capable of supplying fifteen Parisians with their caloric needs—assuming 2,400 kilocalories per person per day, and fifty-four Parisians with their vegetable-based protein of two ounces (fifty-four grams) per person per day. This system provided each Parisian with 110 pounds (50 kilograms) of fresh salads, vegetables, and fruits per person per year. Stanhill acknowledges that this was admittedly an extreme vegetarian diet, but as an urban food-production system, it was "equal to that of the most productive of current, agricultural cropping systems."[2] Furthermore, "in terms of produce yields, the production of the 'marais system' one hundred years ago equaled that of all but the highest-yielding sugar and cereal crops grown today."[3] Stanhill also notes that these market gardeners were not interested in the highest yielding or the most calorie-dense crops. Instead, they were forcing out-of-season and therefore high-profit, quality crops, which makes this growing system even more impressive.

Stanhill was able to account for ten to twenty different salad, vegetable, and fruit crops produced in the Parisian market gardens. *Maraîchers* intercropped; that is, they used companion plants for structural support (planting climbing beans together with corn, for example), or they planted simultaneous sowings of plants known to enhance the growth or protection of another, such as aromatic herbs to keep aphids off artichokes or tomatoes. In this way, *maraîchers* allegedly could reap up to six harvests each year—and never fewer than three. It is thought to be one of the most productive cultivation systems ever documented.

By the close of the nineteenth century, the almost complete replacement of horses by cars as transportation and competition for land inside the city caused a precipitous decline in this type of urban agriculture in Paris. The main ingredient for the raised beds—horse manure—became scarce. Moreover, mechanized transportation allowed for farmers to seek land outside the city and still bring their produce to market.

Urban-intensive market gardening, however, persisted for a very long time in Paris, but it has all but died out in the past two decades. In 1988, there were still 1,900 *maraîchers* growing and selling in the greater Paris region known as the Île-de-France. By 1997, there were 800.[4]

There are now only a handful of urban market gardeners remaining in the greater Paris area, yet I found one quite by accident. While browsing a local street market, le Marché Auguste-Blanqui, with my friend Jill, a Canadian expat who has lived in Paris for over a decade, Jill casually pointed out the market stall of Maraîcher Earl-Raehm, a local urban market gardening enterprise. The table was piled with several types of pointy-tipped red heirloom tomatoes, as well as delicate golden tomatoes that were slightly flattened at each end. There was a pyramid of celeriac root, various types of salads, basil, incredible bunches of

Italian flat-leaf parsley, freshly dug carrots, and potatoes on display, as well. The freshness and the quality were unparalleled. As it happens, this stall is one of Jill's regular stops at this incredible, massive street market that sets up every Tuesday, Friday, and Sunday just a few blocks from her apartment. One look at the expression on my face, and Jill shot me an I-told-you-so smile that said, "Why would we go to the supermarket—in a car—when this is just outside our door?" We spent several minutes mulling over the choices before settling on delicate butterhead lettuce and some endives.

PARIS TODAY

As in most cities in Europe and North America, interest in urban food gardening in Paris hit an all-time low in the 1990s but started to rebound just as it was threatening to become extinct. In 1999, a group of "guerilla gardeners"—activists who plant food gardens on underused or abandoned urban sites without approval of the land's owners—planted an illegal garden on a former industrial site. The project, called the Green Hand, received official approval a couple years later when Paris mayor Bertrand Delanoë supported urban revitalization and urban greening initiatives like Paris's famous Vélib' bicycle-sharing program and the citywide ban on horticultural pesticide use after his election in 2001. Now, La Main Verte is the city's official community-gardening resource organization, and community food gardens are making a come-back to the capital. (In the last decade, there has been a renewed interest in the protection of local culinary traditions, so heritage produce and fruits—Pontoise cabbage, Montmagny dandelion, Argenteuil asparagus, Montmorency cherry, and the Faro apple—are back in vogue.[5]) The city of Paris's official municipal website listed fifty-eight community gardening sites in 2011.[6]

After the morning market trip to Marché Auguste-Blanqui, Jill, her six-year-old daughter, Jesse, and I set out to find a community garden in her neighborhood that her husband, Luc, had stumbled across just a few weeks earlier.

The *Jardins familiaux du boulevard de l'Hôpital* ("community gardens on Boulevard d'Hopital") is squeezed between a 1960s French government–subsidized housing apartment block on one side and high-rent apartments on the other and is accessible only by a sidewalk that cuts between the two buildings. As we approached, we noticed a wiry, gray-haired man fiddling with a row of grapevines, bifocals sliding toward the end of his nose and an unlit cigarette dangling from his bottom lip. His crew neck sweater, worn à la Jacques Cousteau, had a few pulled threads. He could have looked more French only if he had been wearing a beret and had a baguette tucked under his arm as he pruned his vines.

All that protected the garden plots from the footpath was a knee-high fence, with an even lower gate, good for keeping out toddlers or small pets, at the most. He immediately waved us into the garden. He seemed barely interested in my explanation of who I was and why we were interested in this garden. Jean Griffault introduced himself, warmly welcoming the three of us to "his" remarkable garden.

When I commented on the rather ineffective fencing, Griffault conceded that it did little to prevent produce-napping.[7] It was a problem at first, he explained, but it had slowed considerably. It's mostly young kids now, which is normal behavior, he added. He did the same thing as a kid. On the other hand, when grown-ups do it, it *is* annoying. And yes, he has seen adults arrive with shopping bags and proceed to pick anything that looked ripe. But usually there's a gardener around to curb that type of unwanted activity.

"There are twenty-seven plots here," he told us, his unlit cigarette continuously bobbing on his bottom lip as he talked. Every gardener has to look after his or her own space, he stressed. (Gardens of the communal type, where everyone shares in the work and shares in the produce, are popular in France and are often referred to as community

The *Jardins familiaux du boulevard de l'Hôpital* community garden, Thirteenth Arrondissement, Paris, France. Photo by author, October 2, 2010.

gardens, whereas the same term in North America almost always refers to the type of garden where each gardener has a designated spot to work.) Each gardener pays €90 (about $130) per year as a fee for a personal piece of the community garden and for use of the shed and communal tools and water.

Griffault said that he had been gardening here for five years. Before then, he'd never as much as watered a houseplant. "I was born in concrete, and I will die in concrete," he declared rather enthusiastically, explaining that he had been born and lived in the very same Paris neighborhood his whole life. He learned to garden only when he got his plot, mostly by watching the other gardeners.

As we walked slowly through his little garden, he tested our knowledge in a type of name-that-plant agricultural quiz show. The radishes, a bay leaf tree, tomatoes, leeks, artichokes, celery, and strawberries were easy enough. He then moved on to more challenging plants, like lovage and cinnamon basil. His fearlessness in his gardening was endearing—as was his row of Chasselas grapes, though the 2010 humid conditions made it impossible to grow them without being affected by moldy fungus. Nearby, he pointed to a *pêche de vigne* tree, a late-ripening peach, on a neighboring plot. Griffault told us that these peaches were traditionally planted among the grapevines as snacks for grape pickers during harvest.

For a Parisian-born Frenchman, Griffault's sense of international culinary adventure was also impressive. He pulled a long, white two-pound daikon radish (the kind that gets grated into strings and piled on sushi plates in Japanese restaurants), wiped the sticky clay from it, and handed it to Jesse, who really didn't know what to make of it. He also had shiso, a spicy, floral Japanese basil, growing on his plot.

Some plots were like a United Nations of herbs, fruits, flowers, and vegetables. An Antillean gardener had a chayote squash vine. Another had a stand of giant cabbage on remarkably long stocks. We spotted a pumpkin, too, definitely a nontraditional French food. And very late-bearing strawberries—it was the first weekend of October.

"Nipple of Venus!" Griffault shouted naughtily as we approached a tomato plant with purplish-red tomatoes and slightly pointed tips. "It's a new one we're trying this year." At this point, we were clearly pillaging from other gardeners' plots. "Don't worry, I'm allowed," he reassured us, waving his cigarette-holding hand over his head. As we walked, he picked whatever was ripe and handed it to Jesse, who was happily filling her cloth market bag.

After an hour, we knew which arrondissement he had been born in, that he had a girlfriend nearby, that he was crazy about herbs, and that

Community gardener Jean Griffault tends his urban grapevines at
Jardins familiaux du boulevard de l'Hôpital community garden,
Thirteenth Arrondissement, Paris, France.
Photo by author, October 2, 2010.

he was most proud of his compost system and worm bin. Sure enough, as we rounded the garden's toolshed and neatly labeled compost bins, he lifted a strip of burlap inside a long wooden trough, and there were thousands of red wigglers waiting for their next meal of kitchen peelings and compost.

Worried that we'd outstayed our welcome, we thanked Monsieur Griffault for his time—and for our bulging market bag of free produce. Then he invited us to the annual community garden picnic, right there in the garden, the following day. As we walked away, arms straining from the weight of our free Paris-grown urban produce, Jill remarked on the impromptu guided tour, the exchange of e-mail addresses, the gifts of food, the open invitation for a return visit anytime, and the general *bonhomie* of it all. "Happens all the time," I told her. "Whenever little green food plants are involved."

THE KING'S VEGETABLE GARDEN, VERSAILLES

Unfortunately, we couldn't attend Monsieur Griffault's community garden picnic because Jill, Jesse, and I were making our way to Versailles. While the *maraîchers* of central Paris have mostly faded into history, a particularly beautiful example of walled gardening and intensive urban agricultural techniques can still be experienced at the *potager du Roi*, or the King's Vegetable Garden, in this wealthy suburb of the capital, just a few blocks away from the Palace of Versailles.

When Louis XIV, the Sun King, moved his court from Paris to Versailles in the last half of the seventeenth century, a twenty-five-acre (nine-hectare) vegetable and fruit garden was built to supply his kitchen and his lavish parties. No expense was spared. Jean-Baptiste La Quintinie, a former lawyer turned garden designer as well as a highly skilled gardener and orchardist, was in charge of building this garden beginning in 1678. La Quintinie would be the site's head gardener until he died in 1688, and the *potager du Roi* would be his masterpiece.

The site had its challenges. First, a large swamp was drained and dried out. Soil was brought in and stone and masonry walls and terraces were built. The result was a large rectangular central vegetable garden made up of sixteen smaller squares surrounding a large fountain and pond right in the middle. Twenty-nine walled orchards were built around the perimeter. These sunken gardens with high, thick stone walls created microclimates and sheltered areas that allowed fruit to ripen early; even fig trees within this walled, well-tended garden could survive in the central European climate. Louis XIV was said to be fanatical about figs.

Using methods he had learned from *maraîchers*, such as heavy applications of fresh manure from the royal stables, making the most of angles of exposure to the sun, and building thick walls around orchards, La Quintinie was able to ripen fruit five and even six weeks ahead of schedule. Apparently, his gardens yielded strawberries by the end of

March, peas in April, and figs in June. He would even get asparagus and lettuce crops in December.

Despite the drastic reductions in operating funds under King Louis XV, La Quintinie's successors managed to introduce new growing techniques. Twelve coffee plants were eventually acclimatized in the garden's greenhouses, and Louis XV would proudly boast of his coffee grown right at Versailles. The introduction of heated greenhouses yielded even more exotics, most notably the eight hundred pineapple plants growing there when the French Revolution took place.

The garden survived the political and social changes brought on by the revolution and two world wars. The garden's custodians managed in one way or another to continuously produce food, and it now remains largely as it was built over 330 years ago. The *potager du Roi* was listed as a historic site in 1926 and remains one of the few palace gardens left in Europe.

I exchanged a few brief e-mails with Antoine Jacobsohn, the *responsable du service Potager du Roi, Ecole nationale supérieure du paysage à Versailles*—essentially, the head gardener at the King's Vegetable Garden at the National School for Landscape Architecture at Versailles. Jacobsohn, who was born in New Jersey but who has lived in France since the mid-1980s, is equally comfortable with English and French. The garden, he told me, was open to the public Tuesdays through Sundays, from April to the end of October, but on the very weekend that I was in Paris, the annual harvest festival, *Saveurs du Potager*—Tastes from the Vegetable Garden—was going to be held.

After a five-minute walk from the main train stop in Versailles, Jill, Jesse, and I each paid our €4.50 (roughly $6.50) entrance fees.

We had arranged to meet Jacobsohn beside the statue of La Quintinie on the terrace looking out over the central square, which included four hundred different vegetable varieties and five thousand fruit trees (about four hundred varieties). Though little has changed since Louis XIV stood on this same terrace to revel in the splendor of his

Vegetable rows ringed by espaliered fruit trees at the King's Vegetable Garden, Versailles. Photo by author, October 3, 2010.

garden, Jacobsohn immediately clarified that this was *not* a museum garden but very much a working food-producing space, with little time or room for sentimentality.

Varieties don't get to stay in the garden for historic record; they remain if they produce desirable fruit or vegetables that taste good or are interesting in some sort of way for educational or research purposes. If a new cultivar or variety tastes or produces better, Jacobsohn lets the laws of natural selection take effect. "Just because it's old, doesn't mean it's good," Jacobsohn said.[8]

This is also a teaching garden. "I'm a gardener of gardeners," Jacobsohn joked. It's his job to "grow" gardeners, since the garden functions chiefly through funding from the Department of Agriculture in exchange for the educational programs for the 160 landscape architecture students who cycle through the garden. Shockingly, the garden operates on a meager budget of €40,000 annually. "This is a 9.4 hectare site—about twenty-two acres, operating with ten permanent gardeners," Jacobsohn informed us, as he raised his dark, unruly eyebrows. "We're an extremely prestigious space, and we're also an extremely poor space."

Produce from the King's Garden—forty tons of fruit and twenty-five tons of vegetables annually—is for sale at the weekly farmers' market in the cathedral square across from the garden's entrance, as well as in the retail shop inside the garden. There's also a small income from the entrance fee to the garden. "We try to make up for [lack of money] through passion and good organization," laughed Jacobsohn, who was then pressed back into official duty to lead another public tour of the garden on this very busy, crisp fall afternoon.

Left to meander on our own through the various outdoor "rooms" in this enormous garden, we were particularly taken with the orchards. Surrounded by the largest collection of espaliered fruit trees in the world, I marveled at the strange and inventive pruning techniques on display. Espaliered trees (or plants) are those that are pruned specifically into flat, two-dimensional planes, often against a wall or in formal patterns. It's a practice that goes back to at least the Middle Ages and was popular for fruit trees so that they could grow flat against a stone wall and benefit from the maximum amount of heat radiating from the wall, so that the fruit could ripen faster. There are other forms of espaliered trees that are not two-dimensional but are still intended to let as much sunlight hit the fruit as possible, such as the *Doyenné du Comice* pear tree, which was being grown in the shape of a vase on a cylindrical cage for support.

The apple trees had already been picked, but the pear trees were heavy with beautiful, smooth-skinned pears, with provocative varietal names like *Cuisse-Madame*, which translates to "a thigh of a woman," or the more pious *Bon-Chrétien* ("Good Christian")—allegedly the favorite pear variety of La Quintinie himself, and known as William's pear in the United Kingdom, and as Bartlett pear in North America. We found a ripe pear, of undetermined nomenclature, within arm's reach, so I discreetly picked it as a treat for the train ride back to Paris.

In penance, I bought several books at the retail shop, a souvenir kitchen apron for a friend, and an 8-ounce (250-gram) jar of *Potager du Roi* syrupy golden estate honey, from the summer 2010 flow, to add to my ever-expanding collection of urban honey.

PARIS'S URBAN VINEYARDS

Paris's urban vineyards, which up until the eighteenth century covered vast areas of the city, have withstood the onslaught of urbanization better than the market gardens. There are currently 132 urban and peri-urban vineyards in the greater Paris region, and ten inside the city limits.[9] Some are comprised of new plantings, less than two decades old, others date back to the 1930s. All but one are co-ops or municipal plantings that mark major historic vineyard sites in the capital. Le Domaine de la Ville de Paris-Bagatelle, which borders the Bois de Boulogne, is the one private commercial vineyard inside Paris.

Parc Georges-Brassens, just east of the Eiffel Tower, is a vineyard of seven hundred Pinot Noir vines planted in 1983. It's on a site that once held a major vineyard covering the entire south hamlet of Vaugirard up until the end of the eighteenth century. Perlette and Pinot Meunier grapes were added after 1983, as were some Chasselas de Fontainebleau table grapes. Depending on the year, this urban vineyard yields up to 1,300 pounds (600 kilograms) of grapes from a 1,300-square-foot (1,200-square-meter) footprint.[10]

In the thirteenth century, monks tended a thirty-seven-acre (fifteen-hectare) vineyard in east Paris, some of which is in part now Parc de Belleville in the Twentieth Arrondissement. It was reestablished in 1992 as the 2,700-square-foot (250-square meter) vineyard at the Parc de Belleville. It contains 140 vines, some Pinot Meunier and some Chardonnay.[11]

The Vigne du Parc de Bercy, on the Quai de Bercy in the Twelfth Arrondissement, lies along the Seine River. Wine grapes were thought to

have been planted here over two millennia ago when the Romans con-
quered Gaul and named the settlement Lutecia (it was renamed Paris in
the third century CE). By the thirteenth century, Paris had a major vineyard and winery on this site, and Louis XIV had established a wine ware-
housing enterprise at Bercy, which flourished until the 1950s. In 1996, a
total of 350 Sauvignon and Chardonnay vines were replanted, as well as
a number of historic table grapes.[12]

The oldest vines in Paris are now in the Vigne du Clos Montmartre,
planted in 1932. It's also the largest urban vineyard in the city at 16,750
square feet (1,556 square meters). On a good year, the twenty-seven different wine varietals, 75 percent of which are Gamay, yield 2,200 pounds
(1 metric ton) of wine grapes. Le Vieux Montmartre, a historic (and viti-
cultural) society, tends the vineyard, along with the city's parks department. The grapes are harvested and made into wine. The harvest is
marked each year with a public festival in early October.[13]

This wouldn't be the end of my fascination with urban vineyards.
Paris is not the only city in Europe with notable amounts of urban land
under vine. Vienna is the king of urban viticulture with 1,700 acres
planted, producing excellent wines within the city limits. Viennese wine is
best known for its floral whites like Grüner Veltliner, Riesling, and a grape
called Gelber Muskateller from the aromatic muscat family. London,
however, not Vienna, was my next stop. I was thrilled to learn that a few
optimists in London, as it turned out, were starting to grow grapes, and
that London wine, quality notwithstanding, was making a splash.

Paris Nature Capitale

A friend's jogging route takes him past the Arc de Triomphe in central
Paris each morning. On a May weekend run, he was unexpectedly lost
despite this daily routine. He came out onto the Champs-Élysées, Paris's
famous boulevard, and found himself in a forest. Besides the forest, there

**Champs-Élysées, Paris, Nature Capitale—a creation by Gad Weil,
May 22 to 24, 2010. Photo by Nature Capitale/Resolute D.R.
Used with permission.**

were also squares of waist-high green wheat and stands of sunflowers, sugarcane, and bamboo swaying in the morning breeze. Other little plots were dotted with cabbage, lavender, and tobacco. Sheep, cattle, and Limousin pigs were in pens, taking it all in. For as far as he could see, the Champs-Élysées—both the busiest thoroughfare in Paris and home to its most expensive commercial real estate (and therefore the most expensive boutiques and shops)—had become a strip of vegetation, livestock pens, and curious human onlookers.

My friend had stumbled upon Paris Nature Capitale, an event staged by French "street artist" Gad Weil to celebrate International Day of Biodiversity during the International Year of Biodiversity in 2010. Through massive organizational feats, the Champs-Élysées was closed to traffic late Saturday evening, and throughout the night, as teams of volunteers coordinated deliveries of 150,000 plants, 8,000 plots of earth, and 650 fully grown trees. By morning, three-quarters of a mile (1.21 kilometers) of the Champs-Élysées had become a pastoral promenade like its Elysian Fields' namesake.

The event was staged to remind Parisians that their food doesn't come from grocery store shelves but from farms. The 55,000-member French Young Farmers union played a major role, but it also required help from truckers, forestry specialists, greenhouse growers, event planners, and volunteers. Monday night, all was whisked away, returning the Champs-Élysées to its usual busy thoroughfare.

The 2010 event attracted 1.9 million people who came to marvel at the difference between triticale and hard spring wheat, to see pineapples growing at the end of a plant, and to see mustard in full yellow bloom. The plants and produce were for sale, and Parisian butchers organized a mass barbecue onsite. The event made the rounds in the news abroad and prompted a flurry of phone calls and e-mails from friends who knew I'd be interested. The event was such a success that weekend, Nature Capitale is planning future expositions of massive "overnight" gardens for streets in New York, London, Milan, Shanghai, and Moscow.

Urban Beekeeping

About a century ago, urban beekeeping wasn't newsworthy. It was common practice to keep bees in cities, as it was on small mixed farms, for personal stores, or to sell commercially. Perhaps because sugar became cheaper and cheaper from the late 1940s on, honey became less prevalent in our diets. City bylaws were passed outlawing beekeeping, labeling the honeybee as a pest rather than the keystone pollinator that it is in the natural environment. Eventually, it was just something that fell out of fashion. Certainly my friend, a young professional beekeeper, had been used to being the only woman in a male retiree's club until a few years ago.

Whatever the reason, urban beekeeping has been taking off in cities like Toronto, New York, Vancouver, and just about everywhere I visited. (I've heard that Tokyo's Ginza district is a hotbed of urban beekeeping.) It's been an especially important element in the urban-agriculture revolution, given the devastating effects of the mass honeybee decline known as colony collapse disorder (CCD), a name given to the mysterious die-off of anywhere from one-third to 90 percent of beekeepers' hives in 2007. Colony collapse disorder isn't what happens when a beekeeper finds unusual amounts of dead bees in the hive; it's when the beekeeper finds near-empty hives despite the presence of a queen and has no clue as to why the bees left the hive and their queen. Colony collapse disorder continues to claim at least one-third of the honeybee population in places like North America, the United Kingdom, and the European Union, to date.[14]

According to the USDA, there were five million "managed hives" (probably commercial hives for honey production and pollination) in the 1940s, but that number has currently fallen to half despite the increased demand for the pollination services of honeybees.[15] While the honeybee population was already in decline, the recent die-offs raised questions about the vulnerability of the food chain if bees were to continue to mysteriously disappear.

The USDA acknowledges the importance of honeybees in our food chain. "One mouthful in three in the diet directly or indirectly benefits from honey bee pollination," it states unequivocally on its Frequently Asked Questions page for CCD.[16] Studies are being carried out all over the world to try to determine the causes. But the signs are now pointing to a "perfect storm" of stresses that the bee colonies are under: epidemic infection rates by the varroa mite parasite, poor nutrition as these commercial pollinators forage on low crops with low nutritional values for bees, heavy use of pesticides on commercial crops that bees work as hired pollinators, "monocrop diets" of commercial pollinator bees who forage on just a few crops rather than on the more nutritionally varied "polycrop diet," limited access to water or contaminated water sources, and migratory stress as they are trucked up and down the continent as fewer bees are asked to do more and more work.

Plainly put, the countryside is toxic. Rural bees continue to decline while their city cousins thrive and produce almost three times as much honey. Earlier blooms (due to the fact that cities are generally a few degrees warmer than the surrounding countryside because of the amount of concrete), much greater biodiversity per square mile, fewer genetically modified plants, and fewer pesticides are just some of the reasons why bees are happier and healthier in cities.

Bees in Paris

A century ago, Paris used to have over a thousand hives in the city.[17] But there were very few, if any, left after World War II, except for the hives maintained at the Luxembourg Garden Beekeeping School in Paris, established in 1856.

Colony collapse disorder has been especially troubling to anyone in the agricultural chain, but especially in France, which is the European Union's biggest agricultural producer. Since 1995, over one hundred thousand hives have been lost. Pollinator bee populations continue to succumb to CCD by about 30 percent a year, yet the root causes still remain uncertain.

In 2005, France's National Apiculture Association launched an urban bee program to support and encourage urban beekeeping. It was created not only to raise awareness and interest in beekeeping among the public, but it also may help to bolster flagging rural honeybee populations, with city bees being used to rebuild the stocks of rural bees.

While the scientists work out the details of what is killing rural bees, a quick look at the healthy and productive hives in a city like Paris might yield answers more quickly. In the 1980s, Jean Paucton, a prop worker at the Palais Garnier, home of the Opéra National de Paris, put a hive on the rooftop there. It was meant to be only a temporary situation until he could relocate the hive to his country house, but he was shocked to find that in no time the hive was full of honey and the bees seemed happy. They were obviously having no trouble finding pollen and nectar (bee food) along central Paris's chestnut tree–lined Champs Élysées or in the gardens of the Palais Royal or even in window boxes kept by apartment dwellers. More hives went up, and Paucton became a cause célèbre, even getting his photo in the pop celebrity and society magazine *Paris Match*. Now France's most famous urban beekeeper maintains five hives on the roof of the Palais Garnier. His "honey harvested from the rooftop of the Paris Opera," as it states on the labels, is also some of the world's most expensive. A small four-ounce (125-gram) jar sells for €15 ($22). At €120 ($175) per kilogram, it's about ten times as expensive as any other premium honey. Nevertheless the 2,500 to 3,000 jars made each year always sell out. He also has since installed bee hives on the ultramodern Paris Opéra de la Bastille and at his country house after all, though he reports that his rural hives produce less honey, and the country bees keep dying off.

Paucton's success both with beekeeping and commercially with his urban honey has encouraged a new wave of urban beekeepers to install hives on rooftops and balconies throughout the city.

In 2006, Corinne Moncelli got the idea of keeping bees on a balcony at her three-star Eiffel Park Hotel, and she now reaps over three hundred pounds of honey a year (a treat for guests) from three hives. They are tended by Parisian beekeepers Michèle Bonnefond and Armand Malvezin. Bonnefond and Malvezin also keep ten personal hives on their penthouse apartment balcony in Paris.

The Grand Palais, one of Paris's most popular attractions, got its first rooftop hive in 2009. There are now up to five hives, which have proved to be not only a great addition to the Grand Palais's gift shop but

Parisian bee-keepers Michèle Bonnefond and Armand Malvezin check their urban beehives on their balcony in Paris. Photo by Eric Tourneret. Used with permission.

a public relations windfall as well. There are now hives on a skyscraper in the La Défense business district; the town hall in the Fourth Arrondissement has rooftop hives; and Charles de Gaulle Airport has four hives as part of its "show of duty to the environment."[18]

Paris, with all its parks and mild climate is a very bee-friendly city. Most importantly (for bees), it has been a pesticide-free zone since 2000. (The irony that cities like Paris can be refuges for bees tells us about the state that we've created in our rural agricultural lands.) Urban bee-keepers must register with the veterinarian authority of Paris, and their hives must be at least eighty-two feet (twenty-five meters) from the nearest school or hospital. Swarms do happen, but Parisians seem rather content to share their city with docile honeybees.

London, United Kingdom

Perhaps London, with its five thousand hives and a thirty-to-one bee-to-human resident ratio takes the prize for the most productive city for urban honey.[19] For many years, Steve Benbow has tended hives on the rooftop of Fortnum & Mason, one of London's oldest, exclusive food shops right in the heart of the city. Benbow's bees forage on the forty-two acres of private gardens of nearby Buckingham Palace, so it's no surprise the honey sells out quickly, even at £12.95 per 8-ounce (227-gram) jar.

Aside from his exclusive hives up at Fortnum & Mason, Benbow keeps hives all over London and reports that the variety of locations affords him honeys whose tastes range from light honeydew with sharp citrusy top notes to dark toffees from other parts of the city.

In December 2010, the first-ever London Bee Summit brought together scientists to talk about the challenges facing the honeybee in Britain and urban beekeepers whose numbers have been doubling in a very short time (and are likely to keep on growing, as the Capital Growth organization launched a Capital Bee campaign in late 2010 for the city to encourage and support community beekeeping). There was even a

London honey-tasting competition at the London Bee Summit. Seventeen different London honeys were entered. The colors of the honeys ranged from a dark molasses to red amber to those as light as champagne. The winning honey was produced at a nonprofit eco-garden and training center for disadvantaged youths called Roots and Shoots.

Toronto, Canada

As it turns out, the crisp white jackets, white chef's hats, and knee-length aprons are ideal beekeeping attire. "You don't want to wear dark colors," my friend David Garcelon, executive chef at the Fairmont Royal York hotel in downtown Toronto, warned me as we climbed the last few flights of stairs to the rooftop herb garden and apiary.[20] Dark-colored clothing, especially brown and black, puts the hive on high alert. It's too "bear-like," he dead-panned. Bears, as even the most citified person knows, are the bees' public enemy number one. "Oh, and bees tend to be cranky in the morning," Garcelon continued. "And don't stand right in front of the hives' entrance," he quickly informed me, as if he were debriefing me to meet a head of state (apparently everyone has this tendency to stand and observe right in front of the small opening at the front of the hive; bees like a clear runway to their home).

The day I had scheduled for my tour of Garcelon's rooftop apiary was already not ideal. It was drizzling, and there was a little bit of wind. "Bees hate wind, too," he informed me. All I needed to do was to douse myself in perfume (perfume doesn't make bees angry, but it would certainly attract unwanted attention from them), and I'd have an urban bee adventure to remember.

In 2008, Garcelon decided that beehives would be a great addition to his rooftop herb garden. With the help of the Toronto Beekeepers Cooperative, three hives were installed.

"The first question was if bees could survive fourteen stories up in a major urban center," Garcelon told me, as I foraged a few late-season alpine strawberries from nearby raised garden beds also on the four-teenth-story roof. The rooftop herb garden, with its seventeen beds of lavender, thyme, rosemary, chives, strawberries, cape gooseberries, toma-toes, basil, mints, and other culinary plants would provide some suste-nance for the bees. But fourteen stories up, with the wind and the height, and without any other obvious food sources, no one knew if the bees would make it.

As it turned out, the bees did just fine. Garcelon was able to harvest 380 pounds (172 kilograms) of honey—that's over 100 pounds (45 kilo-grams) of honey per hive—that first year. As we looked out over the parapet toward Lake Ontario, Garcelon suggested that maybe the bees forage on a small chain of little islands, just opposite downtown in Lake Ontario known as the "Toronto Islands," or, perhaps they went as far as the Don Valley, a slash of greenery over three miles away.

In 2009, Garcelon and his beekeeping team doubled the number of hives to six, but it was a cold, wet summer. The harvest was a mere 425 pounds (193 kilograms). He'd just harvested and bottled honey two weeks before my visit. As he handed me the jar, he warned that it wasn't a good idea to open it on the roof near the hives. Honey is food for bees, so they're pretty interested in it, and an open jar of honey is just a bad idea. Of course, all these rules are common sense when you think about them. But they were rules picked up along the way, sometimes the hard way. An early-morning television media event to announce the hotel's rooftop hives concluded with a reporter wanting to taste some honey on the roof but ended with him running off-screen on live morning television.

Safely indoors, I cracked open the jar of urban Toronto honey. The late-season honey was a beautiful translucent amber color, tasting a bit like buttery caramel or toffee. Not surprisingly, the menthols and spici-ness of lavender and mint were apparent in the Fairmont Royal York hotel honey. It was also rather thick and sticky.

One of the joys of honey, I discovered as I traveled around and sought out "city honey" from places like Paris, Versailles, London, and

Vancouver, was the different seasonal tastes and *terroir* expressed in the jars. By *terroir*, I mean that the unique characteristics of honey (or wine for that matter) are influenced by the interplay of the weather that year, the specifics of the geography of the place, the season it was made and harvested, and the inherent geological attributes that play out in the year's production. The small pot of Versailles honey, for instance, was thinner and clearer, almost oily, but in a delicious "clarified butter" kind of way. It, too, had mint and lavender top notes but without the caramel undertones of the honey I got in Toronto. Vancouver honey was intensely floral and sweet, characteristics that most likely came from the apple blossoms in the nearby rooftop herb garden and orchard. They were a snapshot of the variety and flavors that the bees had sampled, like a season suspended in liquid sunshine.

Another joy of tasting urban honey was the layering of flavors due to the variety of flowers that were encapsulated in each jar. Late-season honeys tended toward these darker, rounder flavors, while early-season honey had higher-pitched sweetness and subtler floral notes. But it's not just the variations caused by the different seasonal blooms; city honeys are known to contain hundreds of types of pollens, whereas rural honeys, especially those near monocrop fields, can contain less than a dozen.

I was becoming an urban honey snob.

Chapter 7

LONDON
Capital Growth

Cities, like people, are what they eat.
—Carolyn Steel, *Hungry City:*
How Food Shapes Our Lives, 2009

Russians are the European leaders in household food growing. The country's Union of Gardeners reports that seventy million Russians own food-growing spaces. Sixty-five percent of Moscow households grow some of their own food.[1] St. Petersburg urbanites aren't far behind, with 50 percent of households in a city of almost four and a half million growing food.[2] The historical reasons range from frequent interruptions in the food system to the fact that a centralized economy and an overwhelmingly industrialized, chemical-heavy agricultural scenario—even in postcommunist Russia—means that if you want organic, fresh produce, then you likely have to grow it yourself. But very little of that produce is urban agriculture. Instead, household food gardening happens almost exclusively in plots and at country homes, or dachas, outside the city. It's not uncommon for a family, no matter what its economic class, to have a dacha, since Russia has a mushrooming surplus of rural land as the population shrinks and concentrates in cities. These country-garden plots can range from a winterized country house to a glorified garden shed with a bed.

There are recent reports, however, of community gardens starting in cities like Moscow, modeled on Western European and North American ones, where a Muscovite can rent a plot for one thousand rubles (about thirty-three dollars) a month.[3]

Berlin is another city steeped in a long history of personal-use food gardening, but here, it is right in the city. Berlin has 540 "colonies," which are large urban spaces divided into eighty thousand privately owned allotment gardens, and a century-and-a-half tradition of urban food growing. Even Albert Einstein had an allotment in Berlin, but he was notified that he was going to forfeit it unless he started to properly maintain his unruly plot.[4] (Take note, messy gardeners, this could be a sign of genius.) Berliners can even have their mail redirected to their allotments in the summer.

However, I was drawn to Bristol and London because, food-wise, the United Kingdom and North America were in the same proverbial sinking boat of almost complete reliance on an industrial food system with a supermarket-retail model. Britons eat the most "industrialized" diet of any European nation—one in three meals is a type of fast or "ready-made" food.[5] One-quarter of the population is obese[6]; childhood obesity is skyrocketing; and Britons waste a tremendous amount of food through the industrially supplied supermarket system: 6.7 million metric tons a year, or one-third of the food purchased in each household.[7] Britons import most of their food from outside the country, despite having a solid agricultural sector, and they are as utterly dependent as we North Americans are on the hope that the shelves of the supermarkets will continue to be full. In other words, city dwellers in the United Kingdom are in the same precarious position of experiencing alarmingly high levels of food insecurity. In fact, it was Lord Cameron of Dillington, the then head of the UK Countryside Agency, who declared that the United Kingdom was "nine meals from anarchy" after an in-depth 2007 report on the nation's food system was conducted.[8] Maybe it was the wake-up call that British cities needed, since food security and urban agriculture finally seem to be getting some traction in places like London and Bristol, and interesting things are percolating as a result.

London not only has a history of fervent urban food gardeners growing on some thirty thousand allotment plots, but right now, huge swaths of the city are under construction or scheduled for redevelopment. London's mayor, the kooky, two-wheeling, scarecrow-haired Boris Johnson, has an aggressive and overt green agenda. (His daughter's middle names are "Lettice" and "Peaches.") I had read about green-thumbed entrepreneurs, like London's newest (and only) commercial vineyard owner and vintner, whose newly planted vines are growing in an industrial area of the city. The city-supported Capital Growth Network had announced a goal of helping London establish 2,012 new food-growing spaces by 2012. And the 2012 London Olympic Games committee had announced that it had a food policy for the games—the first of its kind—that would leave a legacy of a strengthened local food system in its wake.

The United Kingdom was always a bit ahead of the curve anyway. And London was the city that ushered in the Industrial Revolution, so maybe it was poised to lead a type of deindustrialized urban food revolution.

ST. WERBURGH'S CITY FARM, BRISTOL

Unlike in North America, where developments and roadways are given pastoral names for marketing purposes, Watercress Road in Bristol is actually named for the natural watercress beds that could be found there until a century ago. (The cress beds gave way not to the farm but to housing in 1910.) The houses were destroyed in an explosion nearby as a storm drain was being built. The derelict houses were finally torn down in the 1970s, and in 1980 area residents rehabilitated the site as a city farm with the support of Bristol City Council. It was Bristol's second city farm.

Graffiti next to signage for St. Werburgh's City Farm and Café, Bristol, United Kingdom. Photo by author, September 29, 2010.

City farms seem like vestiges of village life in the United Kingdom but are, in fact, later arrivals on the urban-agriculture scene. They came out of the 1970s in London and spread to Bristol. Windmill Hill Farm was Bristol's first city farm in 1976, then came St. Werburgh's in 1980. City farms are exactly as they sound, small farm holdings in urban areas that contribute to the social fabric of the immediate community. They provide a rural space, food, food education, and a community green space for urbanites.

Prince Charles has been the patron of the Federation of City Farms and Community Gardens, the United Kingdom's main city farm umbrella organization, since 2001. There are nine city farms within the inner metropolitan limits and sixteen within the Greater London area. The oldest is the Kentish Town City Farm, founded in 1972, right in the heart of Camden, in King's Cross, inner-city London. According to the Federation of City Farms and Community Gardens, which supports 120 or so community-managed affiliate city farms, there are a thousand community gardens and orchards, a growing number of affiliate allotment gardens, and two-hundred-plus city farms and community gardens in development—city farms in the United Kingdom are a £40 million ($63 million) industry. Very few charge admission. Instead, they depend on private and public funding as well as income from therapeutic social horticulture and agriculture programs that benefit socially disadvantaged youth at risk of slipping into crime and poverty and physically and mentally challenged clients.

St. Werburgh's City Farm, perhaps because its creation predates the founding of the Federation of City Farms and Community Gardens, operates independently as a separate registered charity. It is both a working farm—raising pigs, goats, sheep, ducks, chickens, and rabbits for meat—and a greenhouse and training center for youth and mentally challenged persons. The public is welcome, free of charge, to wander through and observe the animals and take in the different rhythms of a small mixed farm, which includes a community garden and greenhouses in a highly urbanized area, like in the heart of Bristol. The main purpose of the farm is for people to have a place where they can experience a bit of rural living regardless of their income or physical capabilities. As such, it's open almost year-round, free of charge.

"We run it as close to a working farm as we can," said farm manager Tim Child, on a drizzly, concrete-gray morning when I showed up for a tour of this urban farmyard.[9] Already soft-spoken and measured with his words, he proceeded with caution, "Everything we produce here, we use for meat." And then he looked for my reaction. Clearly, he had had his run-ins with vegetarian journalists.

A sow just had a litter a week before, so we scrambled over a wood fence and into a concrete shed. The largest sow I had ever seen in my life was lying on her side, sleeping with a large hairy ear shielding her eye from the light. Thirteen piglets were piled into a corner under the orange glow of a heat lamp.

Still measuring my ability to stomach such issues, Child mentioned that he intended to keep six piglets to fatten but the rest would be sold off right away. Some might even end up on the menu at the nearby farm café, he offered. I didn't fake bravado, but I didn't pretend I didn't eat bacon, pork chops, or ham either. Once he realized that I wasn't going to break down at the thought of the piglets becoming sausage meat, he continued. When the piglets were weaned, the sow would go, too. "As soon as I decide I'm not going to breed them again, they get replaced," he said.

When Child came to the farm, there were too many old animals, he said, and the place just wasn't functioning very well. Nor was it very interesting. Having worked on a commercial hog farm with 3,500 pigs, he had to figure out how to keep a good variety of animals and have good turnover on such a small site. And the solution was that the animals raised on the farm could not become pets.

Finding slaughtering facilities willing to take on one or a few animals at a time, Child said, was problematic for small mixed farms in the United Kingdom. Luckily, he had a deal with a slaughtering plant that was only fifteen miles (twenty-four kilometers) away. Selling meat was

Food-production poly-tunnel at St. Werburgh's City Farm,
Bristol, United Kingdom. Photo by author, September 29, 2010.

never a problem. He kept a list of area residents who would buy as much as he could sell, and there was also the farm's independently run onsite café. Demand, it seemed, always outstripped the farm's ability to supply the local neighborhood.

I asked if customers pay more for his locally produced meat, and Child replied that they pay a "fair price." "They've seen where the meat comes from; it's fresh, and they also want to support the farm."[10]

Meat sales, however, were not the main income source for St. Werburgh's. Charitable donations from private and public sources were the mainstay of the farm. It also helped that the farm paid only a "peppercorn rent," a very low fee, in other words, to the Bristol city council, owners of the land. Plant sales from the onsite greenhouse also brought in a few pounds here and there, and there was income generated from the training programs that the city farm provides to "clients." (St. Werburgh's provides horticultural training to adults with learning disabilities and also offers youth programs to troubled school-aged youth.)

Spotted pigs, St. Werburgh's City Farm, Bristol, United Kingdom.
Photo by author, September 29, 2010.

"I didn't expect that I'd be in a job where fifty percent of my time is working with adults with learning disabilities," said Child. A lot of those clients with learning disabilities come in as troubled youth, kicked out of school and with serious social integration problems.

"You get a report on these lads from school and it's best not to read it," Child confided, though I was sure his imposing, well-muscled volunteer firefighter frame would give even the most rebellious client cause to think twice about acting up. But according to Child, he's never had a problem with one of the kids. "You get them out of that environment [where they were getting into trouble]. You get them away from their peers, and they have no one to play up to. They become totally different."

After chatting with Child, I wandered the farm's pathways with moms pushing strollers and children racing between the various pens of sheep, chickens, ducks, pigs, and other animals enjoying the cool, misty day. When the mist turned to rain, I popped into St. Werburgh's Farm Café at the edge of the farmyard. Part circular tree house, the café attracted a bohemian crowd. In the middle of the restaurant, there was a large table of what looked like at least three sets of Gor-Tex®–clad parents with their toddlers. The adults drank coffee and laughed as the children grazed plates of farm-sourced eggs, sausages, and seasonal vegetables. But for all its quirkiness, St. Werburgh's City Farm Café had a serious national reputation, as did its workhorse thirty-something chef, Leona Williamson, for the café's simple, straightforward farm-style menu and for Williamson's commitment to sustainability and community-minded social values. Instead of listing the miles of its food on its menu (which it doesn't even bother to do), this cafe could rightly list food yards. The pigs that were reared on the farm could also end up on the café's menu.

When the downpour returned to a drizzle, I set out for the asphalt walking and bicycle path to the Ashley Vale Allotment Association gardens. I was curious to see an example of Britain's famous allotment gardens.

While allotment gardening may seem like a leisure activity to many these days, its history in the United Kingdom (and almost universally) began as a social concession to the landless poor. With the political and social upheavals of the British Civil War through the seventeenth century and continuing with the enclosure of common land in the eighteenth century—which dispossessed the farming poor, leading to increased urbanization in the nineteenth century—the allowance for allotments as a civil right got some political lip service with the Allotment Act of 1887. The Victorian idea that gardening would be a way to keep the poor away from drink and other activities resulting from idle hands was really just a way to try to keep poor people from causing trouble, not as a result of social justice values redressing inequality. As such, the legislation had little traction.

The allotment concept finally got the legislation it needed starting with the Small Holdings and Allotment Act of 1907 and 1908. This legislation, and further amendments, gave the movement stricter guidelines and legal force. If six or more people demanded land for growing food, the town council had to provide it. From the end of the 1800s to the First World War, allotment plots in the United Kingdom soared from a quarter of a million to one and a half million.

The Dig for Victory campaign sponsored by Britain's Ministry of Agriculture during World War II seemed to put a positive spin on the hardships brought on by the war and the legacy of food rationing, which lasted in Britain into 1954. The Ministry of Agriculture estimated that in 1941, nearly 1.3 million tons of food was produced on allotments.

This is roughly the same ratio of production to space under spade that was confirmed by a 2008–2009 survey. The standard 300-square-yard (250-square-meter) plot produces an average of three-quarters of a ton of food. At 330,000 plots, this means that the United Kingdom currently produces 247,500 tons of food. Yet with over one hundred thousand gardeners waiting for allotments, it could be 322,500 tons. Given that a standard plot can provide a family of four with "a reasonable portion" of its annual fruit and vegetable needs, 430,000 allotment plots could provide 107,500 families of four with their fresh produce per year.[11]

This Dig for Victory campaign traveled oversees to North America, where "Victory Gardens" were set up all over the United States and Canada during the Second World War. While in Britain, the aim was to keep civilians from starving during the deprivations of war; the rhetoric in the United States enlisted gardeners into the fight more directly. I still have my grandmother's book *Victory Backyard Gardens: Simple Rules for Growing Your Own Vegetables with Simple Rules and Charts*, which was published in 1942. The book begins with an address by the secretary of agriculture, Claude R. Wickard. "One indispensable line of war production is food," Wickard asserted. "Let's make it the three V's—Vegetables for Vitality for Victory," he concluded.[12]

Allotments, however, declined in Britain (as did the vacant-lot gardening and other wartime urban-agriculture measures), as they did throughout Europe and North America. By 1970, allotments had dipped to around half a million plots in the United Kingdom, giving way to the pressure to use the land for other urban development priorities. The allotments that remained were increasingly poorly kept. And by 1999, there were only an estimated 250,000 allotments in Britain.[13]

But allotment demand has roared back into prominence in Britain during the past decade. Currently, waiting lists for allotment plots can have more than a thousand applicants in some city council areas. The National Society of Allotment & Leisure Gardeners (NSALG), the main national representative body for the allotment movement in the United Kingdom, reported that wait-lists grew by another 20 percent in 2010.[14] The most recent figures from the society estimate that a minimum of 330,000 allotment plots are currently in use in the United Kingdom. There has been a feverish increase in demand, with over a thousand people currently on waiting lists. The society noted that one strategy was to cut the size of allotments in half—as is common in London—for new allotment gardeners as they wait for full-sized plots.

The same press release detailed the benefits of allotment gardening. The first benefit listed was healthy exercise. The second was good company. "Quality, fresh, and affordable fruit and veg" is the third bullet on the list. While exercise and good company is undoubtedly a concern for people who face food insecurity, it might not be their top priority. A 2006 article by the BBC titled "Can You Dig It?" that came out a week in advance of National Allotment Week (August) had already identified this trend. "Younger professionals" are "swelling the ranks of the allotment diggers." The article also mentions that allotments are considered "green gyms."[15] (One can only expect that the allotment workout DVD will be next, if it doesn't already exist.)

Even at the allotment garden's entrance, wet earth, compost, and the pungency of celery and root vegetables hung in the air. Many of the community garden spaces I have visited in Canada and the United States suffered from over-manicured neatness—more eye-candy than geared toward food production. The gardens in England's Ashley Vale Allotments, in contrast, were refreshingly wild, messy, unruly, and, as a consequence, quite productive. Being the end of September, allotment owners were clearly struggling to keep up with the fall productivity of most of the plots. A few gardeners had even turned over their plots to new plugs of winter kale and cauliflower.

The bulk of the plots were long, rectangular beds running lengthwise on a steep vertical slope that ran toward a row of houses. Some had little wooden sheds; others had small private greenhouse buildings. Most had a round, black plastic compost collector. Perhaps because of the rain that day, there were no gardeners for me to chat with about their plots. I was curious to know why the vertical strips ran lengthwise on the slope, as opposed to horizontally—rice paddy–like. I wanted to know what crops they grew and how often they visited their gardens. Was the fencing intended to keep out hungry humans or other grazing wildlife? Instead, I pulled off a few ripe blackberries that found themselves on the outside of the metal fencing and deeply inhaled the smell of wet earth and plants.

Ashley Vale Allotments, founded in 1917, is home to 214 allotments of varying sizes that cover a 13-acre (5.3-hectare) parcel
near St. Werburgh's City Farm, Bristol, United Kingdom.
Photo by author, September 29, 2010.

VERTICAL VEG: FOOD GROWING FOR SMALL SPACES

My first appointment in London was to meet Mark Ridsdill Smith, a
North London food gardener whose nonprofit enterprise Vertical Veg
"inspires and supports food growing in small urban spaces." We had
traded e-mails when I stumbled across his blog on the web, and when I
announced that I was heading to London, Ridsdill Smith graciously
invited me over to tour his balcony gardens.

It wasn't difficult to spot Ridsdill Smith's house from the street. It
was the one with giant gourds and tomatoes cascading from second-
story window boxes on an otherwise confusingly repetitive array of late
Victorian, veggie-less row houses.

Ridsdill Smith, his wife, and their toddler live on the top two floors
of the house, which means they don't have access to the backyard below.

But Ridsdill Smith had always wanted to grow food, so he had signed up for allotment space with his local council in 2000. After seven years of waiting, he checked back with them. "I was told that it would be another fifteen years, so I'd be about sixty-five years old by the time I got one."[16]

Concerned about climate change, Ridsdill Smith had been working to make an impact with a number of "low-carbon groups," but he found it was tricky to get people enthusiastic about all the things they couldn't do. He turned to growing food, figuring it would be a low-impact action that might be a nice change because it was about something you *could* do. He had always kept a few planters of arugula, salad greens, and herbs, so he wondered just how much he could grow using the space he had available to him: a nine-foot-by-six-foot northwest-facing balcony, six windowsills (four facing south, two facing north), and a small patch of concrete outside the front door. As it turned out, the answer was *a lot*.

Ridsdill Smith began documenting the production he was getting on his Vertical Veg website and set up a social enterprise nonprofit business of the same name. He needed a benchmark challenge and wondered if he could make up for his lack of allotment space with just his available balcony and windowsills. Could he grow at least £500 ($820) of produce that he might otherwise have to buy at a store in a year? He documented each handful of parsley and each harvest of tomatoes, peas, and beans, month by month, itemizing what he'd otherwise have to pay at a greengrocer. By the time I got to London in late September to meet with him, he'd surpassed his initial goal and was shooting for a new target. He revised his goal to aim for the equivalent of £782 from May 2010 to May 2011—the average annual crop value that comes off a London-sized allotment. (The National Society of Leisure & Allotment Gardeners estimates that a 300-square-yard (250-square-meter) allotment produces £1,564 worth of food a year. Most London allotments are half this size.) In the end, during his 2010–2011 May-to-May experiment, Ridsdill Smith's balcony gardens would produce eighty-three kilograms of food, or the equivalent of £899.99, almost double his initial goal.[17]

The bulk of Ridsdill Smith's food growing was taking place on his small wooden balcony above the neighbor's backyard below. This space—so small that Ridsdill Smith and Clare Gilbert (a student apprenticing with him during a year off from her academic studies) had to take turns on the balcony—was home to an amazing selection of food: lemongrass stalks, Bright Lights Swiss chard, various types of tomatoes, a bay leaf tree, rosemary, salad crops, kales, blueberries, lovage (a stalky and leafy herb with a flavor similar to strong celery), and even a mallow, the highly versatile family of plants that includes marshmallow, okra, and cocoa. The tower of scarlet runner beans yielded about 2.2 pounds (1 kilogram) per week at peak production. A wooden window box had a new crop of sunflowers and beans that would be harvested as sprouts. Pots, two and three deep and arranged in a ring around the outside perimeter of the balcony, had an unruly display of edible plants. And a small plywood box housed Smith's wormery that kept the plants in a

View of the ultra-productive nine-foot-by-six-foot balcony of Mark Ridsdill Smith, founder of Vertical Veg, London, United Kingdom. Photo by author, September 30, 2010.

steady supply of nutrient-rich compost. Above the patio, a cantilevered wood-plank shelf was loaded with several buckets of lettuces.

No space was wasted. And no culinary experiment, it seemed, was out of the question. Ridsdill Smith lamented that his wasabi, a Japanese horseradish, was struggling, but the daikon radishes that the women from London's legendary Coriander Club (a Bengalese women's community garden that is part of the Spitalfields City Farm in London) suggested he grow were flourishing.

Ridsdill Smith's living room had become a plant nursery. Now that his son was old enough, a diaper-changing table had even become a multi-tiered lettuce nursery. The windowsills that faced the street were crammed with window boxes of herbs, tumbler tomatoes, salad greens, kale, peppers, and newly planted cilantro, just past the two-leaf stage. The same was true for the bedroom windowsills: zucchini, radish sprouts, tomatoes, and peppers. I half-jokingly mentioned that the window boxes looked heavy. Falling window boxes, toppled by a bumper crop of tomatoes, were a real possibility, he replied. He pointed out his extra staking and ties to prevent any disasters. He had also been experimenting with a "Canadian invention," which he'd learned about on the Internet. The window boxes were actually two planters, one set inside the other, double boiler–style; the water in the bottom box would draw up into the soil in the inner box via osmosis. This kept the soil's nutrients from washing out of the soil as they tend to do when watering takes place from above.

Ridsdill Smith was modest about his impressive balcony-based food production. Truly, he made it all look so easy. He then handed me a bag of incredible salad greens. The Thai coriander would end up scenting my shoulder bag for the next few days, but the windowsill produce was eaten that day in my newly acquired addiction to "ploughman's lunch"— the iconic British sandwiches of sharp, hard cheese; relish; fresh or pickled vegetables; and sliced onions for which I had acquired a fondness relatively quickly.

Soon after the tour of his impressive gardens, Ridsdill Smith, Clare Gilbert, and I made our way to a new food-growing space Ridsdill Smith had helped with—a first of its kind in the United Kingdom—a grocery store with a rooftop veggie garden in nearby Crouch End in the London borough of Haringey.

FOOD FROM THE SKY

Crouch End, with its white-accented brickwork townhomes and picturesque 1895-built clock tower, looks like a life-sized gingerbread village. I was grateful that Ridsdill Smith was navigating to the store with the mouthful of a name, Thornton's Budgens of Crouch End.

Thornton's Budgens are well-known grocery stores throughout the United Kingdom. They are smallish, niche grocery franchises where each store has some leeway to specialize in local fresh foods tailored to its customer base. Thornton's Budgens of Crouch End, owned by charismatic Irishman Andrew Thornton, had recently made the food pages as the first Thornton's Budgens in the chain to sell gray squirrel. (The United Kingdom was still undecided at the time as to whether the resurgence of gray squirrel as a meat option was the latest in sustainable locavore eating or a stomach-churning animal rights travesty. Crouch End apparently had a pro-squirrel-meat consumer base: a July 2010 article that appeared in the *Guardian* newspaper reported that Thornton's Budgens' Crouch End store was selling "10 to 15 a week" when they could keep it in stock.[18])

Ridsdill Smith and I weren't there for the squirrel meat or even the retail experience, however. We were there to see "the world's first supermarket roof garden."[19] Ridsdill Smith had been one of the twenty volunteers who transformed a supermarket's flat roof into a 4,800-square-foot (450-square-meter) food garden in the spring of 2010, complete with scarecrows, Tibetan prayer flags, and a hammock.

As we arrived up on the rooftop, Food from the Sky's creator, Azul-Valérie Thomé, was squatting on a low stool, rubbing mustard greens seeds out of dried pods of a row that had been left to seed. It was a "seed day," according to the biodynamic calendar, she told me, extending her hand, with a playful acrylic ladybug ring on it, to greet me.

Thomé told me that she had been "studying Steiner," referencing Rudolph Steiner, the father of biodynamic agricultural practices, for some time.[20] Biodynamic food growing is an organic, holistic approach to farming and food production, incorporating celestial calendar days for optimal seeding, transplanting, and harvesting. While it seems a bit mystical, I've learned not to dismiss biodynamics' seemingly kooky concepts, because I'm most often incredibly impressed by its healthy crops and tasty results.

After living on a fifty-acre organic farm in the county of Devon in southwest England for fifteen years, Thomé arrived in London with "this vision of wanting to cover all the flat roofs with orchards and strawberries," she explained as she continued to hull seeds, her accent a seductive layering of British over French. Thomé informed me that there are about one hundred square kilometers, or one billion square feet, of roof space in the city. Shortly after arriving in London, she met Thornton. "He said, 'I have a roof. Want to come up and see it?'" And Food from the Sky was born in spring 2010.

Ridsdill Smith gamely took over seed collecting as Thomé walked me around the rooftop farm. Some of the seeds, Thomé explained, would be given away free or donated to local community groups or individuals. Some would be saved for future plantings. Some would be "returned" to the Heritage Seed Library, a UK-based "lending library" that provides various endangered seeds, mostly European varieties, that have fallen out of favor with commercial seed companies or that were never available in commercial catalogs to begin with. Food from the Sky was playing its part by keeping these rare vegetable variety seeds in the hands of growers, to keep the biodiversity and community food resilience from further erosion, and as a way to help build up community reserves of these open-pollinator seeds—plants that pollinate in the wind or that exist thanks to visits from bees or other flower-loving insects—for future generations.

Rather than spreading soil over the roof (which generally means some expensive weight-bearing assessments, reengineering, drainage installation, and waterproofing), hundreds of plastic recycling bins were donated by the Haringey council, and ten tons of LondonWaste—organic compost that London collects as household kitchen and garden waste, composts, and then redistributes to backyard gardeners, allotment owners, and anyone looking for growing medium—was delivered to the roof. (The city's household green-waste collection program diverts 45,000 metric tons (49,604 tons) of organic waste away from landfills every year.[21] Methane is a gas that is 21 percent more harmful environmentally than carbon dioxide; therefore, diverting large amounts of compostable matter plays a huge role in sustainable urban planning.)

Thomé then coordinated volunteers to plant, transplant, nurture, and harvest the variety of crops that grow astonishingly well in their plastic recycling bins. Thomé excitedly pulled a baseball-sized Bull's Blood beetroot for me to photograph. It was growing happily in a recycling container next to a fig tree that sported several small figs dangling from the thin branches. A straw-mat vertical fence offered protection to the row of tomato vines and arugula, radish, and various other seeds that were being sprouted in repurposed cardboard egg cartons.

I remarked on the abundance of bees working the flowers. Their buzz was audible even above the street noise. Thomé said that the bees arrived just one week after the first blossoms appeared on the plants. One of her volunteers has identified up to thirty different insects on the roof. Pollinators like bumblebees, honeybees, butterflies, and solitary bees clearly enjoyed the buffet of pollens at their disposal. But spiders, flies, and worms were also noted as a sign of a healthy ecosystem. "It was dead a few months ago, a very desolate urban area. Life has come. They've moved in," Thomé boasted, letting out a throaty laugh.

**Bull's Blood beet, grown on the roof of Thornton's Budgens
supermarket at Food from the Sky, London, United Kingdom.
Photo by author, September 30, 2010.**

Seed gathering gave way to harvesting, cleaning, and packaging
romaine heads. (By the time I left, crisp bullets of romaine hearts from
the roof were on sale in the special display island at the front of the
store.) Depending on the season that summer, Crouch End customers
had been able to purchase salad greens, arugula, choy, cress, big black
radishes for the winter, purple broccoli, borage (a culinary and medic-

inal herb with star-shaped blue, white, or pink blossoms), tomatoes, parsley, beets, carrots, and sorrel (a perennial leafy herb for soups and salads), just hours after they had been picked.

Despite her products getting premium prices thanks to the quality and freshness, Thomé noted that the income required to keep Food from the Sky operational could not come only from food sales. Workshops and educational visits "giving opportunity to children and older people to show them how to grow food in their homes and how to increase urban biodiversity" were part of the plan. Then she flashed a smile and said, "Biodiversity includes humans too, you know." Food from the Sky had a higher purpose than providing ultra-fresh lettuces and vegetables to affluent Crouch End gourmands. "Between permaculture and biodynamics, we feel that we are connecting the whole of life, of creation really, in 450 square meters. This is an amazing opportunity. Not only just to grow food, but to create a template of what's possible on a roof."[22]

FROM URBAN BLIGHT TO URBAN WINE

When I mentioned to friends who had lived in London at one point or another that a commercial vineyard had been planted in King's Cross in 2009, there were two reactions: confusion and disgust. (King's Cross's reputation for being a gritty, postindustrial, and particularly unsavory part of central London won't last long though. It's undergoing a massive revitalization.) The thought of British wine alone was unthinkable, they told me. But London wine? Absolutely, if novelty counts for more than drinkability. London has had an underground urban wine movement for a couple of years, and I was determined to see its first commercial winery for myself.

My enthusiasm was slipping into foreboding, however, as I made my way up Camley Street from the King's Cross Underground subway station. The area was one giant construction site, and apart from the sparkling new St. Pancras Eurostar Chunnel train station, King's Cross had a long way to go toward the promise of urban renewal. About twenty minutes along, I finally reached a mixed residential-industrial neighborhood, where I was hoping that I wasn't searching in vain for a vineyard.

By this point in my urban-agriculture adventures, I had learned to keep an open mind. The location rarely resembled any of my preconceptions. In this case, the vineyard, as it would turn out, would be tucked around the back of the two-story Alara Wholefoods warehouse building.

Dressed in a slate-gray suit with a claret-colored tie, Alara owner Alex Smith looked less like a vintner and more like a CEO. But as we walked and toured his "dream farm," as he calls it, I began to see that he was not your average business owner and that this was not your average warehouse head office.

Rather than heading to the "vineyard" right away, Smith wanted to show off his food gardens. More specifically, we started at his garden's toolshed/*cave de vin*, a canary-yellow repurposed shipping container at the edge of a garden pathway. Truthfully, it *had* been a long, taxing walk from the train station to Alara, so when Smith offered me a taste of his estate elderflower "champagne," I didn't say no.

It was tart, sharp, and refreshing, almost grapefruity. We were literally standing under the canopy of the century-old elderberry tree whose small white spring flowers and wild yeasts made that tipple. Well, I thought, if you can make a drinkable sparkler from an old tree on an industrial site, maybe there is a vineyard hidden somewhere here. We sipped from our flutes as Smith told me the story of how he came to be London's first commercial vintner.

Alex Smith came to central London to study architecture in the early 1970s.[23] He quickly became alarmed at the indecent profits that developers were making off of certain areas of London, while many were finding that the cost of living in the city meant squeezing more and more people into poverty. Smith became concerned with the fate of Tolmers Square, an early-Victorian landmark at the north end of Tottenham Court Road, a major commercial road in central London. Tolmers Square was next on the redevelopment auction block, and shell companies were in the habit of buying up property, intentionally vandalizing them to make them unlivable, and then turning them into huge redevelopment properties. When one of these intentionally derelicted properties crumbled, killing two pedestrians on the sidewalk, Smith decided that he had to take a stand.

"There I was studying architecture and this sort of thing was going on in the profession I wanted to go into," Smith explained as we nursed our flutes of sparkling elderberry.[24] Smith despaired both at the disregard for safety and at the thought of the vandalism to a beautiful historic square just in the name of profit. "The only way I could really properly morally oppose it was to live there without money," Smith said, referring to his decision to become an illegal squatter in this area slated for redevelopment. He also began a year of living entirely without money. (When he received a £5 note as a birthday gift that year, he used it to start a fire for warmth.)

The yearlong project came to an end when Smith found a £2 bill on the street. He borrowed a friend's truck and used the money to pay the vehicle entrance fee to the New Covent Garden Market, an enormous vegetable and fruit market in southwest London. Smith went around to the vendors' garbage bins and gleaned unsold and unsellable fruits and vegetables. Back at his squat, he sold his findings and made £5. He then used the profits to return to the market, gleaning more unwanted foods to sell to his squatter community. The next week he bought some wholegrain flour at the market and used it to bake bread in an oven left behind in the squatters' residence. He also bought other whole foods like rice and dried beans for resale. Alara Wholefoods was born.

Smith and his wife, with whom he founded and still runs the company, moved Alara several times, most recently in 2005 to a patch just near the Eurostar railway tracks in King's Cross north of St. Pancras station. The site cleanup was epic, according to Smith. "Fifty tons of rubbish" had to be removed. Fencing was installed and terraces dug in for gardens. "If you can imagine the sorts of detritus you'd get in a dark corner within walking distance of a train station," said Smith, trailing off to let me fill in the imagery. "It was rather insalubrious here," he finally added.[25]

That said, clearing the land yielded a few nice surprises. One large pre-Roman building stone was unearthed, as were two large, flat Roman slabs and a Georgian cornerstone. As it turns out, Alara sits on land in King's Cross formerly known as Battle Bridge. This ancient crossing of the Fleet River is where the Romans apparently fought a game-changing battle around 60 CE with the Iceni, a Celtic rebel army led by warrior queen Boudicca that resisted Roman occupation of "Londinium." The Romans won, and that was the end of the Celtic resistance, though there's now a huge bronze statue of Boudicca in her war chariot with her daughters on the north bank of the Thames near the Parliament Building. There's even a discredited urban legend that persists that Boudicca is said to be buried under Platform 10 at King's Cross train station. As Smith recounted this bit of King's Cross history and smatterings of urban folklore, I scribbled "PR bonanza!" in my notebook and figured marketing companies should be falling over each other with a winery backstory like this. Moreover, I marveled at the fact that this Eden-like garden sprouted on a spot where a millennium of constant human occupation could be traced.

Yet pineapple guava trees thrived. As did pomegranate trees. Blackberry and raspberry canes ran amok. There was also waist-high rosemary, sage, feverfew (a medicinal flowering herb), and comfrey (an important organic fertilizer). There were Asian pear trees, goji berries (also known as wolfberries—red, elongated berries that grow on hedges), asparagus ferns, apricot trees, blueberries, kiwi vines, tomatoes, globe artichokes, pears, gooseberry bushes, broad beans, and French beans. And there was something Smith called a blue bean tree, a perennial bean tree with periwinkle pods. There were over fifty types of edibles growing in this garden adjacent to a warehouse.

It was more than enough for staff lunches, sourced from the garden year-round, Smith explained. Schoolkids also came for programs where they toured the gardens and then picked and prepared lunch in the warehouse kitchen.

In 2009, Smith planted a dozen fruit trees on a neighboring business's boulevard adjacent to the commercial parking lot. In a few years, Booker Wholesale, the discount grocery "cash and carry" will be able to offer a bonus of free tree fruits as an added incentive to shop there.

Finally, we got around to the side of the warehouse where the vineyard was located. I realized that the 2009 article in the *London Evening Standard* that had referred to the vineyard as "London's first large-scale vineyard" was typically hyperbolic in its media pronouncement, but it was an urban vineyard nonetheless—and London's first commercial one in quite some time.[26]

The vineyard covered a berm that ran 148 feet long and twenty-six feet wide (forty-five meters by eight meters) and had thirty-some Rondo vines—which offer an early-ripening, frost-hardy, and downy mildew-resistant red hybrid grape—that were planted in February of 2009.[27]

By my visit in early October, the 2010 harvest had already taken place. Although the vines were still very young, Smith said that he'd harvested thirty-five kilograms (seventy-seven pounds) of grapes, which should yield about twenty-five bottles of wine. This was a trial vintage, and the juice was still happily fermenting away at Smith's house. The official commercial vintage would be available in 2011, and he'd employ the expertise of London's Urban Wine Company for the first official commercial harvest, crush, and bottling. Smith already had a buyer for some of the first vintage, which he hoped would be around a hundred bottles. Acorn House Restaurant, near the vineyard, had plans to put Alara's 2011 wine on its wine list.

Alex Smith, urban vintner,
Alara Vineyard at Alara Wholefoods,
London, United Kingdom.
Photo by author, September 28, 2010.

LONDON'S URBAN WINE COMPANY

Pursuant to this idea of urban grape growing in London, I followed up on Alex Smith's comment about London's Urban Wine Company. Richard Sharp, a South Londoner, came up with the idea for the Urban Wine Company while vacationing in France. He marveled at how entire villages came together during harvest to pick and crush grapes. He had a few vines in his garden at home, as did a number of friends in his borough of Tooting in south central London. He suggested that they pool their grapes, and the first press was in the fall of 2007. The Urban Wine Company was born. Its first vintage was twenty bottles of Chateau Tooting, Furzedown Blush.[28]

It is now well known that cities, when they are large enough and sufficiently concretized, create their own microclimate. This is known as the "heat island" effect. The United States Environmental Protection Agency (EPA) defines a heat island as when "built up areas are hotter than nearby rural areas."[29] Furthermore, the EPA notes that the annual mean air temperature of a city with one million people or more can be 1.8–5.4°F (1–3°C) warmer than its surroundings. In the evening, the difference can be as high as 22°F (12°C).[30]

It was actually in London in the early 1800s that this phenomenon was first described by an amateur meteorologist named Luke Howard and was dubbed the heat island effect. Howard noticed that it was several degrees warmer at night in London than it was in the surrounding farmland areas. Currently London, with its 7.83 million residents in the Greater London area and with a surrounding metropolitan area, supports a population of twelve to fourteen million residents.[31] This means that in central London, it is often five degrees Celsius (nine degrees Fahrenheit) and as much as ten degrees Celsius (eighteen degrees Fahrenheit) warmer than its surrounding rural areas.[32] This gives London a growing advantage that can compete with Germany and northern France.[33]

With all the allotment and backyard gardeners in London, there are a surprising number of grapevines snaking their way throughout the boroughs. In 2010, the Urban Wine Company had one hundred Londoners in its cooperative. The stipulation for membership was a minimum of three kilograms of ripe, healthy grapes, delivered to a collection point on a specified "Harvest Day." (Earlier-ripening grapes are to be picked when ripe—when the sugar-acid ratio is good—and frozen, if need be.) The grapes are trucked to a winery on the outskirts of London, de-stemmed, crushed, and placed into a fermentation tank for a few months, then bottled. The harvest resulted in 1,300 bottles of London wine.[34] Each cooperative member received six bottles in return. With the minimum requirement of about one ton of grapes to make a separate red or white, the wines so far have been made from the combination of both red and white grapes. But with a harvest of 1.5 tons (3,306 pounds) in 2009, and another 1.5 tons in 2010, it's possible that soon the tonnage will be at levels high enough to produce separate whites, reds, and even sparkling wines. According to Sharp, grape-growing communities across London are blossoming. In May 2010, the Urban Wine Company even planted

Urban Wine Company founder Richard Sharp with a bottle of 2008 Chateau Tooting Furzedown Blush, London, United Kingdom. Photo reprinted with permission of Urban Wine Company.

thirty vines at Eastlea Community School in East London, presumably to nurture the next generation of London vintners.[35]

London's Food Games 2012

Wandering south along Midlands Road after my visit to Alara Vineyard, I came upon King's Cross Skip Garden—fortunately, but quite by accident. The space is essentially a construction brown site that has been turned into a food garden, thanks to some inventive reuse ideas with huge metal construction rubbish bins, or skips.

King's Cross is home to London's largest redevelopment site. Some sixty-seven acres (twenty-seven hectares) are slated for redevelopment by 2020 in the King's Cross Central master redevelopment plan. Construction cranes and job-site fencing were the dominant aesthetic, so food gardens were the last things I expected to find in this mega-construction site. But this, as it turned out, was one of the sites encouraged and supported by Capital Growth, the city-supported green initiative whose goal is to create 2,012 new food-growing sites in London by 2012.[36]

I peered through the cutouts in the artfully decorated hoarding that wrapped around the Skip Garden long enough that a young, jeans-and-T-shirt-clad man waved me in. Bert Dutka, a twenty-something program animator with Global Generation, had just finished with a class of young schoolchildren who had spent the afternoon at the Skip Garden learning how to grow food in urban spaces even if they don't have access to a garden or allotment.

Skips are mobile industrial construction wastebins used in the United Kingdom, and they are everywhere in this massive construction zone. They are used mainly to contain and remove construction waste from industrial building sites. As it turns out, they also make handy, and portable (if you have access to a crane) food-growing spaces. The Skip Garden was created in the summer of 2009.

Dutka works with Global Generation, a youth-based organization dedicated to empowering youth with the knowledge and skills to effect change toward more sustainable communities. Global Generation maintains the Skip Gardens and runs the educational programs. Dutka said that around four hundred kids a year attend programs at the Skip Garden site.[37] They get to see how food plants grow, what they look like, and, moreover, they get hands-on experience with planting, looking after, and harvesting food.

Three skips in the middle of the yard are for annuals: parsnips, broad beans, red and green chard, beets, herbs, corn, with some nasturtiums and other flowers on the borders. One of the skips has a polyethylene-covered half-tunnel, the skip garden's rendition of a greenhouse.

One of the skips along the side of the lot is a portable orchard: five

Inside the Skip Garden, King's Cross. St. Pancras Eurostar station is on the left, and construction skips repurposed as mobile food gardens are on the right. London, United Kingdom. Photo by author, September 28, 2010.

apple trees, two pear trees, two grapevines, raspberry canes, and other soft fruit such as cascading alpine strawberries. The orchard skip is a demonstration of a permanent planting scenario, teaching kids the difference between annual plants, perennials, and permanent plants.

The portability of the skips is important, said Dutka, since Global Generation has until 2020 to use the skips and move them from construction site to construction site as the rebuilding of King's Cross Central takes place. The idea is to plant the fruit trees and vegetables on the rooftops of the new buildings in the area as the redevelopment progresses.

Paul Richens, a rooftop gardens entrepreneur and site manager for a number of the urban-agriculture sites, including the King's Cross Skip Garden, chimed in that he was drawn to the project because, as he put it, "this is an exercise in gardening in difficult places."[38]

"This is quintessentially urban 'ag.' You don't get any more urban than this. It's about using a space which is unusable," Richens said, as we walked the uneven packed ground that was a combination of construction dust and rubble.

Richens was most proud of "the green engine," the skip that held three separate wormeries, something he called a "worm café." A lush strip of Russian comfrey, the unimpressive-looking plant that is a very rich source of nitrogen and that makes excellent organic fertilizer when pulped or soaked in water to make "compost tea," ran lengthwise on the lip of the green engine. This wouldn't be the first time I saw a comfrey juicer. The four feet of PVC drainpipe were held vertically onto the side of the skip. The idea was to pack freshly cut leaves into the pipe and crush them with a weight on the end of a chain, the dark-green sluice emptying into a modified plastic gallon milk jug. "And the most disgusting-smelling liquid in the universe comes out the bottom," enthused Richens. Comfrey, as any gardener will tell you, smells like hell but is the best organic fertilizer going.

Richens has his sights set on a secondhand 1,000-litre (264-gallon) in-flight fluid container, the kind used on aircraft. He said he needed three hundred liters (eighty gallons) a week, on average, for watering. And when a heat wave hits London, as it did in the summer of 2010 with its record high temperatures that went up to 116 degrees Fahrenheit (46.4 Celsius), Richens went through twice that much. London tap water was neither environmentally savvy nor desirable, because it didn't have the right pH for the plants. And he wasn't keen on the chlorine in the tap water either. "If it kills bacteria on our teeth, what else is it killing in the soil?"[39]

TASTIER, HEALTHIER, GREENER: THE LONDON OLYMPIC GAMES OF 2012

Perhaps London's most ambitious local food initiative to date will be the London Organizing Committee of the Olympic and Paralympic Games (LOCOG), which will take place in the summer of 2012.[40] LOCOG is the first Olympic Games organizer to put forward a food vision as part of its organizational plan and bid. The LOCOG official document, "For Starters," refers to its vision as "aspirational standards"; it will set goals for sourcing and supply chain for its commercial partners, caterers, and suppliers. Because of the intensity involved in feeding the athletes, the visitors, and the support staff, it's still a move forward to use a major international event to fast-track more local, more sustainable food systems in London and throughout the United Kingdom. (In 2000, Londoners consumed 6.9 million tons of food. Eighty-one percent of that was imported from outside the country.)[41]

What this means is that the athletes' villages, the main press center, the International Broadcast Center, and the overall food that the thirty-one competition venues at the 955 competitive events—estimated at some fourteen million meals—will fall in line with a food policy. It will put British-produced food at the center of attention of the visitors and athletes. Imported foods such as bananas, tea, coffee, sugar must be Fairtrade (certified by Fairtrade International) products, meaning they must meet the committee's standards for ensuring that the producers receive a living wage for their products. Chocolate must be either Fairtrade or ethically sourced—that is, products grown or produced in ways and conditions that are nonexploitative to the workers, animals, or environment where they are made. Traditional British cheeses—for instance, cheddar—must be made in Britain. There will be an estimated 25,000 loaves of bread, 232 tons of potatoes, 75,000 liters (19,800 gallons) of milk, 19 tons of eggs, and 330 tons of fruit and vegetables consumed in the Olympic Village during the games. To achieve these numbers, and to source the products within the United Kingdom, urban food production will have to play a role.

But the Olympics will come and go like a circus through town. Whether the edible window dressing remains afterward is anyone's guess. But one thing tends to hold true: when you habituate people to growing food in a city, it becomes the norm. For all of London mayor Boris Johnson's eccentricities, his government's focus on sustainable food security in London is hard to knock.

Chapter 8

SOUTHERN CALIFORNIA AND LOS ANGELES

A Tale of Two Farms

The Western Land, nervous under the beginning change.
—John Steinbeck, *Grapes of Wrath* (1939)

As I flew into Los Angeles, California, on a clear June day, I tried to map out the flats, farm belts, and orchard lands that drew the great caravans of migrants west during the Dust Bowl years, as well as the giant commercial enterprises that now lure people from Mexico and Central America to work the strawberry and tomato fields, to pick oranges, and to harvest avocados. But the approach over the greater Los Angeles area became a geography lesson of another sort. The continuous urban landscape where 12.8 million people[1] (and their cars) were living was without natural reference points. As the airplane circled and descended, the city became an enormous monochromatic computer circuit board with an infinite horizon. Its features were distinguished only by varying shades of gray: lighter gray boxes and rectangles that made up warehouses, shopping malls, and parking lots; the medium gray of the great freeways that cut above and through the city; and the charcoal gray of the smog that sat over what presumably was residential Los Angeles that day. I sat in awe of the lack of green space, despite how much I'd read about the city's complete concretization and continuous urban sprawl.

139

Just north of Los Angeles, across a small mountain range, is Bakersfield, the southern tip of California's great Central Valley. It runs north to south, stretching 450 miles (720 kilometers), and is forty to sixty miles (sixty to one hundred kilometers) wide. This is where the lettuce, tomatoes, and strawberries that fill grocery stores in the United States and Canada—and maybe elsewhere—are grown. Home to one of the great tracks of farmland on the North American continent, the Central Valley is the produce basket of North America, providing 8 percent of the nation's agricultural output on less than 1 percent of the farmland. The mountains north of Los Angeles and those that divide the Pacific coastal cities of Monterey and San Francisco offer a very distinct barrier between rural farmland and the gleaming modernist cities and boundless suburbs that still shape Southern California and, most notably, Los Angeles.

This friction between food and freeways has taken its toll. Los Angeles's metropolitan area, the globe's fourteenth-largest megacity, is home to the very rich and the very poor, a place of great excess and of endemic food insecurity. White neighborhoods in Los Angeles have three times as many supermarkets as do black neighborhoods and twice as many as do Latino ones.[2] South Los Angeles is often held up as a textbook example of a food desert.

Despite the fact that south and central California are extremely productive agriculturally, one million Angelinos go hungry or are food insecure every day, making hunger every bit as much of a chronic condition as the disproportionately high rates of obesity, diabetes, heart problems, and other health conditions that arise when urbanization meets poverty. Los Angeles, according to the Supplemental Nutrition Assistance Program (SNAP; formerly known as the food stamps program) is the "epicenter of hunger" in the United States.

My main purpose in traveling to Los Angeles was to meet up with some of the farmers who got the world's attention as they fought against the destruction of their fourteen-acre urban farm in South Los Angeles in 2004. It was the largest community gardening project of its time, and its destruction became the subject of a heart-wrenching Academy Award–nominated documentary called *The Garden* (2008). Perhaps this urban farm was ahead of its time, but several years earlier, another early urban-agriculture battle took place in the Goleta Valley, near Santa Barbara. Fairview Gardens Farm, which started out as a rural farm but found itself surrounded on all sides by residential sprawl throughout the years, retained its right to exist and is now the Center for Urban Agriculture at Fairview Gardens.

What I found in the Los Angeles area—an early battleground where important struggles in urban agriculture have been fought and won, and fought and lost—is that it is still a frontier in many ways. It's a cautionary tale of how greed, politics, racial disharmony, and extreme urbanization can take their toll on a city's soul. But the city is making great strides to meet these challenges. There are a handful of people working to put food-growing spaces and alternative food-distribution models at the center of its makeover. Food gardens are rising from Los Angeles's notorious food desserts. The city's Unified School District now has five hundred school gardens at varying levels of production. There are seventy community gardens in Los Angeles County, giving food-growing access to 3,900 families. And in 2010, the city put together a high-powered Food Policy Task Force that conducted an inventory and assessment of the city's foodshed and produced a 108-page document in support of a municipal and regional food policy.

A LITTLE HISTORY

Los Angeles County was, until the 1950s, the largest agricultural county in the United States. It produced citrus, walnuts, strawberries, tomatoes, dairy, and meat in significant quantities, along with a wide array of other agricultural products.[3] As its cities grew in the postwar boom, urbanization stampeded through the state. Farming was expected to make way for development. Urban growth—residential communities, schools, retail centers, roadways, commercial zones, industrial areas—seemed to have a presumptive de facto right of way. Food production, farmers, and farmland simply were expected to move out of the way of "progress."

Two notable exceptions to this rule became causes célèbres in the urban-agriculture sphere: Michael Ableman's successful fight to save Fairview Gardens Farms in the Goleta Valley near Santa Barbara throughout the 1990s, and a group of Latino campesinos and campesinas now known as the South Central Farmers, who lost their struggle to continue growing avocado and banana trees, beans, chayote squash, corn, yams, and sugarcane right in the middle of an industrial park and gang-ridden area of South Los Angeles in 2006.

Fairview Gardens Farm

Michael Ableman was an accidental urban farmer. As an aspiring photographer and an enthusiastic back-to-the-lander, he started working at an agrarian commune managing a hundred acres of pear and apple trees in Ojai, California, at the age of eighteen. From there, he continued north along California's Pacific coast to manage an avocado and citrus sapling nursery near Santa Barbara. Still in his early twenties, Ableman arrived at Fairview Gardens, one hundred miles north of Los Angeles and ten miles from Santa Barbara, in 1981, to work and lend his organic expertise, primarily grafting orange trees. The farm manager who hired Ableman soon announced that he was taking a leave of absence and asked Ableman to "farm-sit." The manager never returned, and Ableman's two-decade journey on Fairview Gardens farm began.

Fairview Gardens was originally homesteaded in 1895, with topsoil thirty feet deep. By the 1950s, it was part of the patchwork that made up the agricultural quilt of orchards and farmland in the Goleta Valley.

As Ableman writes of his arrival at Fairview Farms in *On Good*

Land: The Autobiography of an Urban Farm, the Goleta Valley was already in a state of agricultural to post-agricultural transition. The Chapman family that owned Fairview Gardens no longer wanted to farm the land and were quite happy to let Ableman run wild with his ideas of what the twelve-and-a-half-acre farm could be. His tinkering and ambition led him to diversify the farm from a few orchards to a wide variety of fruits and vegetables, including peaches, avocados, citrus, artichokes, asparagus, beans, broccoli, lettuce, melons, tomatoes, carrots, and the like. As most commercial farms in Southern California were growing single-crop fields for supermarkets far away, Ableman writes that he was turning Fairview Gardens farm into a "kind of supermarket in itself, diverse and ever changing like an agricultural botanic garden."[4]

As Ableman's agricultural botanic garden developed, detached homes with swimming pools, sidewalks with streetlamps, lawns, cul-de-sacs, and other signposts of suburbia crept ever closer. The maw of generic residential and commercial development outflanked the farm on all sides by the early 1990s, and the newly arrived citified neighbors weren't happy with the tractors, the on-farm store, the roosters, and the piles of compost. Ableman had to continually defend the farm's right to operate, as the landscape around it shifted from rural to urban. Though the farm grew local, convenient, organic, fresh food, it was at odds with the new residents' ideas of what an urban space should look, smell, and sound like. There were cease-and-desist orders filed over the farm's composting, and the roosters' daily dawn ruckus prompted public nuisance complaints, which in turn sparked a media storm as supporters and naysayers clashed. Ableman was spending a lot of time and energy just trying to keep his farm operating as it had been for decades in the face of the juggernaut of suburban sprawl.

Fairview Gardens produced one hundred different types of fruits and vegetables, employed thirty people, and fed some five hundred families through on-farm sales and farmers' markets. The farm was grossing close to one million dollars in sales at its peak.[5] It was a viable economic entity, standing on its own two feet without any municipal or community concessions, though it was being treated as less than one in so many ways.

The farm's land had already been rezoned as residential, with the potential for fifty-two condominium homes. Developers were salivating, while the family that owned the farm had mixed feelings about the land's legacy for the community and also the potential payday for its inheritors. It was a fight that Ableman surely would have lost if not for the fact that Cornelia Chapman, the family's matriarch, obviously harbored an emotional connection to the land as a farm. Ableman ultimately convinced the family that the land should be left as farmland in a trust to give future generations the ability to see where and how food is produced. He also knew that it was important for the family to get a reasonable value for their land. In the end, a compromise was miraculously reached in the gulf between the then current value of the tract as agricultural land and the unthinkable price it would bring as residential properties. Cornelia Chapman eventually pushed a note under Ableman's door with her final figure. The terms were "Seven Hundred and Fifty Thousand Dollars Firm and No Bickering!" spelled out in handwriting.[6]

Ableman managed to raise the asking price and promptly set up a nonprofit corporation, with a board of directors and bylaws, that would steer the future of the land. The Center for Urban Agriculture at Fairview Gardens became one of the first agricultural protected zones in the United States in 1997. A precedent-setting case had been made for urban agriculture. But in the end, it was all a very gentlemanly agreement over land use, the selling price, and the future of a suburban part of Santa Barbara. Now the farm functions as a working enterprise with a produce stand and a Community Supported Agriculture (CSA) program, an educational center, and a national model of urban agriculture and urban farmland preservation.

Every minute in the United States, over an acre of agricultural land is lost to commercial and residential development.[7] We have built our cities next to good farmland for the obvious reasons, but those flat, well-drained open spaces make for easy sprawl. Land is always more valuable as a future shopping mall, because food can be obtained so cheaply from elsewhere. Clearly the odds are never in favor of the small farm, let alone an urban one.

The story of Michael Ableman and Fairview Gardens is a rare example of success in the face of slim odds. But Ableman had the support of a community—wealthy donors, foundation grants, and regular farm customers—and was able to raise the money he needed. He was also lucky that the Chapman family wanted to cash out and get out of the landlord farming business—but didn't want to gouge. I'm sure it didn't feel like it to Ableman at the time, but it was a fairly straightforward transaction among relative social and economic equals.

A group of disenfranchised Angelinos from South Central Los Angeles did not have such advantages as they coaxed guava trees, sugarcane, tomatillos, and cilantro to rise from the Los Angeles concrete. In the end, the South Central Farm's fourteen-acre patch of green wasn't allowed to survive, as it was too much outside the paradigm. The derelict fenced-in space in South Los Angeles serves unintentionally as a cautionary tale and a reminder that racial inequality and the principles of private land ownership versus the common good are still hugely important shapers of the American city.

SOUTH CENTRAL FARM'S UNLIKELY BIRTH

At 3:15 p.m. on April 29, 1992, a jury acquitted four Los Angeles police officers of using excessive force in the roadside beating of Los Angeles resident Rodney King. The beating, which occurred in 1991, was videotaped by a bystander and was a shockingly brutal account of what many Angelinos of color felt was a common occurrence: unchecked racially motivated violence by a corrupt, mostly white Los Angeles Police Department. The footage of King, an African American Angelino, being beaten with batons and kicked while on the ground was broadcast on the television news and written about in newspapers around the globe. When the predominantly white jury acquitted the officers, it was the breaking point in poor and gang-ridden black and Latino neighborhoods, especially the one known as South Central, one of the most economically disadvantaged areas of Los Angeles.

By 3:45 p.m., a mob of three hundred had gathered at the Los Angeles courthouse. By evening, all hell was breaking loose. The overwhelmed police retreated from the areas of the worst violence. There were three-plus days of roaming mobs, beatings, murder, looting, and arson. Los Angeles was imploding. On the fourth day, the army, the National Guard, and the marines arrived with 4,000 heavily armed soldiers to restore order to the city. There were 55 people reported killed; between 2,000 and 4,000 were injured, depending on the source; 12,000 people were arrested; and there was upward of one billion dollars in damage. The episode would later be known as the LA Race Riots of 1992.

Doris Bloch was the director of the Los Angeles Regional Food Bank located in South Los Angeles at the time. When the Department of Agriculture and city asked her what else they could do to help to restore some normalcy to this battered neighborhood, she pointed to the derelict strip of land across the street. It was, by most accounts, a garbage-strewn, rat-infested brown space. The city had acquired this land in 1986 to build a trash incinerator, but the community protested and the incinerator was never built. Bloch proposed starting a community garden as a way to heal the community and to address the issue of hunger and reliance on the food bank by area residents. In 1994, the city granted the LA Regional Food Bank a "revocable permit" to establish a community garden and occupy the land.

With a large Central and South American population in South Los Angeles, and with many more recently arrived Latinos with extensive agricultural experience, "the gardens," as they were known, quickly transformed the blighted eyesore into an unbelievably lush and productive fourteen-

View of the former South Central Garden in Los Angeles.
Photo © Lane Barden Photography. Used with permission.

acre oasis. Not only did it produce food but it functioned as a "third-space" like a town square that many Latinos were familiar with, a social gathering place where news was exchanged and celebrations were held. And it took the place of urban parkland, allowing green space for family picnics and get-togethers. It provided shade and clean air in a city that suffers from severe air-quality issues. Moreover, it was a haven for kids to run through, and for the Latino gardeners to reconnect with their cultural and spiritual roots. Heirloom vegetables, fruits, and herbs prized by the Mexican community were grown here, such as various types of corn, Mexican wild yams, *chayote* (a pear-shaped gourd), banana, avocado, guava, amaranth (a flowering herb with edible seeds and foliage), sapodilla (a Central American tree fruit), *epazote* (a pungent green herb, similar to cilantro) and *hoja santa* (another green herb, whose name translates to "holy leaf" in Spanish).

Devon G. Peña, professor of anthropology at the University of Washington, inventoried the plant life at the South Central Farm and believes that there were between 100 and 150 different plant species growing there, an example of mixed farming that took into account plants for medicinal, nutritional, and spiritual uses, as well as practical farming approaches such as companion planting. (Corn, squash, and beans are referred to as "the three sisters," as the corn provides the stalk that the bean plants can climb up, squash provides a groundcover under the corn, and beans help with moisture retention in the soil and keep weeds from growing and using up valuable soil nutrients. As it turns out, these three crops are also nutritionally symbiotic when eaten together.)

Aerial photos of the garden show how it was a verdant living rectangular island floating in a sea of asphalt, warehouses, and empty streets. But not everybody in the community was happy with it.

Some gardeners would sell some of their excess produce, which gave the impression that the farmers were profiting from their plots. This was likely just a scapegoat for the community friction that was developing as the Latino presence in the community grew and the older, more established African American powerbrokers in the community felt left out (even though it should be noted that the farmers were helped along in their community organizing by members of South Central's African American community, many of whom had decades of grassroots activist and civil rights experience).

Robert Gottlieb, the Henry R. Luce Professor of Urban Environmental Studies at Occidental College and director of the Urban and Environmental Policy Institute in Los Angeles, concludes in his 2007 book *Reinventing Los Angeles: Nature and Community in the Global City* that the South Central Farm had become "a showcase for inner-city food security, urban greening, and a new type of public and community space."[8] Gottlieb adds that "people were suddenly interested in what was going on there." The story got even more complicated as a third player entered the scene. "Perhaps, most importantly, Ralph Horowitz, the owner of the property when it was first taken over by the city, decided he wanted the property back."[9]

The parcel of land had become valuable again because it was a safe, clean, and orderly part of an inner-city community. It was also located on the Alameda corridor, a freight rail yard that ran between the Port of Los Angeles and Long Beach. And the previous owner, claiming that he was entitled to the right to repurchase the land under the original sale agreement in 1986, successfully sued the city for the right to get it back. Neither the decision to sell the land nor the sale price or terms of the sale agreement were publically released until after the deal was done. Unbeknownst to the farmers, the site was quietly resold to Horowitz in a murky agreement. Later, it was discovered that Horowitz essentially was allowed to buy the land back from the city for just over five million dollars, almost the same price he had sold it to the city for some seventeen years earlier.

It can only be assumed that the farmers were expected to acquiesce, that they would not want to bring too much attention to their cause. Instead, the farmers sought legal advice, reorganized their plots among themselves, and, most importantly, started calling themselves the South Central Farmers Feeding Families. It was decided democratically among the farmers that families should be limited to the number of plots they worked as a sign to the outside that this was not a struggle to make money on the land but was viewed as a right to feed themselves and their families.

Being bullied and pushed into a corner was not an unfamiliar feeling to many of them. This time, however, they decided that they were not going to just leave quietly. They were tied to this land emotionally and spiritually. They were going to fight back.

Tezozomoc, whose father had been one of the original gardeners and who taught many of the farmers how to cultivate their land and grow food successfully, took on the role of political organizer, activist, and outside liaison, as did Rufina Juarez, an articulate and indefatigable farm laborer. The farm's story expanded from one having regional interest to reaching a worldwide audience when filmmaker Scott Hamilton Kennedy captured the final days of the farmers' struggle, which ended with SWAT teams in riot gear arresting protesters and farmers, as well as bulldozers razing stands of corn, guava, and avocado trees, in his 2008 documentary *The Garden*. The eighty-minute film captured the birth of urban farming as a political movement in the United States and the galvanizing moment when food-security issues hit the political agenda.

On the morning of June 13, 2006, Los Angeles police in riot gear used chainsaws to cut through the fencing to enforce an eviction notice. Farmers and protesters who didn't go quietly were arrested. Shortly thereafter, bulldozers razed the more than 350 garden plots. It was the most widely publicized showdown at an urban garden.

The land today is once again derelict, unproductive, ugly, and under surveillance to guard against squatters and the return of any of the farmers. (The city established an alternate seven-acre site in Los Angeles for the South Central Farmers, but most found it unacceptable, in part due to its location and in part due to the fact that it was in a power corridor with high-voltage power lines crackling overhead.) Viewed from the road, the original location of the gardens is a strangely barren site hemmed with chain link to prevent the farmers from returning and planting. The grooves of the bulldozers' routes as well as the outlines of the former plots are still visible from above via Google Earth®.

Following the story and watching the surreal news footage of bulldozers pushing over avocado trees and trampling gardening sheds in Hamilton's documentary left me with a sense of bewilderment. Was that really the end of this incredible farm? Were the bulldozers really the final word in a fight that seemed to have so much hope and so much potential? Wasn't this the perfect answer to extreme urban poverty, disenfranchisement, malnutrition, health issues, and cultural displacement that a multicultural city like Los Angeles should have been supporting? This is what I hoped to find out by traveling to Van Nuys in 2010 to meet Tezozomoc.

It had been four years minus a week since the farm's final eviction notice was posted as I sat across from Tezozomoc. In real life, he's a mountain of a man. The long, wild, wavy hair he sported in *The Garden* has been short-cropped and is salted with middle age. First I asked about his single-identifier indigenous Mezoamerican name. "Just Tezozomoc," he replied. "It's a form of de-identification," he added, cracking a sly smile.[10]

I asked him what circumstances brought his family to Los Angeles in the first place. Tezozomoc's family came to Los Angeles in the 1960s, drawn by the promise of better economic prospects, displaced by the attempt to industrialize agriculture that had been devastating traditional subsistence farming communities. "We all come from rural farming people. We've been farming for thousands of years," he explained in describing the Latino population that coalesced around the garden. "We were all pulled here by economics. We're a product of the Green Revolution."

Tezozomoc's father worked as a janitor but became disabled, so he was unable to continue to work. The garden became his passion. "It was a way of reconnecting."

"It never started out as a political statement," Tezozomoc continued. "It started as an evolutionary adaptation. We're hungry. We can't get a job. We're old. If you went to work, as a single mother, someone would have to take care of your kids, and you couldn't afford that."

One of the first ideological battles the South Central Farmers fought was against being called gardeners. "We weren't gardening. We were growing food!" To differentiate themselves from this world of gentry gardeners, the campesinos eventually called themselves the South Central Farmers Feeding Families and became a self-governing organization.

"That was very important because we wanted to highlight the fact that we weren't doing this for fun. We'd rather *not* be doing this. The question of food is not a theoretical construct. It's a matter of life and death."

"You can misunderstand our struggle, because our struggle was defeated because they didn't want us to be an example of 'the common.'" It did, however, kick at the cracks of the status quo and brought to light the barriers that are used to keep the rights and privileges of the "haves" out of reach for economically and socially marginalized people. The garden and the gardeners represented a type of "outsider" behavior that was rarely tolerated in deeply capitalist, imperialist economies like the

United States: the farm did not assign value to the land in terms of its economic ability to produce income, but rather it was valued in noneconomic terms for its social, cultural, and knowledge-keeping capital. Perhaps even more offensive to the prevailing American sensibility, the farm raised the question of whether a nonoccupying owner who held a legal claim to a piece of land really should triumph over productive occupants providing a demonstrable social good.

"What we represented, what South Central represented, was a form of escape . . . an escape from that system of control. Why? Because somehow we managed to get a piece of land. Somehow we managed to make decisions on it. Somehow we managed to sustain life on it."

Tezozomoc wishes that the documentary had been more explicit about what he perceives as the heart of the problem, "the whole question of survivability in highly urbanized spaces," as he put it. "The facade of urbanism is something that we're all going to have to engage with at some point. This is an unsustainable environment any way you slice it."

Concessions were made by the city, and an alternate gardening site was established. Some of the farmers dispersed but many continued to grow food. It also served as a catalyst that sparked other indigenous urban food gardens, and it presented a model for community-based self-rehabilitation and self-reliance through urban organic agriculture.

As I probed further on the subject of food security in Los Angeles, Tezozomoc grew quiet. "I don't know what makes this city tick. There's no industry anymore. Los Angeles is having a hard time maintaining its state of exception from nature." He hinted that the answer to Los Angeles's current crisis may be beyond the scope of urban agriculture. The future is a return to healthy, local, natural food. Urban agriculture puts that discussion front and center, especially in a place like Los Angeles.

Since 2006, the South Central Farmers have continued to grow and bring fresh, organic food into the communities that need it most in the inner city. But they've had to lease land as far as a four-hour drive north of the city in Shafter. Despite the commute, the farmers service eleven farmers' markets, eight of which are in Los Angeles. Ironically, they have also started selling at farmers' markets in the largely commercial agricultural communities of Bakersfield, where it's often just as difficult to access fresh, organic produce as it is in the city. The South Central Farmers are now selling food in two counties with their CSA program.

*   *   *

At the urging of Tezozomoc, I remained in the Los Angeles area for a few more days to attend a dedication celebration of new farmland, a private donation of a scrub-brush-covered eighty-five-acre piece of land in Buttonwillow, two hours north of the city. Tezozomoc hinted at "a big announcement" he was planning on making at the event, which was also timed to mark the four-year anniversary of the eviction from the inner-city farm.

The permanent Buttonwillow farm means that the farmers will never be forced off their land again. It's right in the heart of Kern County agricultural heartland, where massive monoculture plantings of cotton, almonds, potatoes, and other commodity crops are grown. Despite the overwhelmingly "conventional" feel of Buttonwillow, the South Central Farmers, and their CSA model of organic farming, have found themselves welcomed into this community. Several local businesses pulled together to give this cooperative the essential operating component that it was missing. A well and irrigation system worth over $200,000 was donated. This will provide organic produce at affordable prices in the heart of California's agricultural heartland.

Native American elders and dancers of all ages from the Chumash Nation had begun drumming, dancing, and blessing the land at 5:45 a.m. By noon, when I arrived, the wind was already unpleasantly hot, and the only shade was the tent city that ringed a flat piece of field where the dancing, drumming, and singing continued despite the heat. When the dancing ended in mid-afternoon, this signaled the time when the well would be turned on.

Banner marking the new farm for the South Central Farmers in Buttonwillow, California. Photo by author, June 12, 2010.

Congresswoman Maxine Waters, an outspoken and long-time supporter of the South Central Farmers, gave a fiery speech about "land that transformed a whole area from simply a ghetto to a productive piece of property. And then because of greed, and a lack of humanity, the struggle to hold onto that land was a struggle like many of us have not seen for a long time."[11] With that, she marched across the dusty field in her high heels, cut the ribbon on the well, and flipped the switch, starting the pump. Everyone cheered as the first sputterings of murky water shot out from pipes. The very next day, the farmers could begin to plant.

"I don't think you can really do a book about the global food movement without talking about the South Central Farmer," John Quigley told me, over his plate of beans, rice, cilantro and a grilled, stuffed flour tortilla, as we waited out the heat of the day under the tents.[12] Quigley

Chumash dancers began the ceremony to bless the land at sunrise and danced until midday, South Central Farm, Buttonwillow, California. Photo by author, June 12, 2010.

is an environmental activist who specializes in "tree-sits," an act of environmental awareness and civil disobedience carried out by occupying a platform on a tree that is scheduled for destruction. He spent twenty-three days living at South Central Farm and coordinated a rotating roster of Los Angeles celebrities who lent their time and media-grabbing power to the farm. Quigley's role was to bring the celebrity power and media attention to make sure either that the farm was allowed to stay and function or that its destruction did not go unnoticed.

Los Angeles, according to Quigley, has the least amount of green space of any city in the United States. "In the early 90s, we flew in the Goodyear Blimp over LA. Over fifty percent of it is concretized. That's what made 'the farm' even that much more special.

Lunch served up at "Farmchella," the celebration of the new South Central Farmer's land, Buttonwillow, California. Photo by author, July 12, 2010.

South Central Farmers hold baskets of produce grown on the other sites they have been using as they wait for their Buttonwillow Farm to become operational. This produce is sold at farmers' markets in the Los Angeles area and via the South Central Farmers' CSA program. Photo by author, July 12, 2010.

"On a hot day like today on the street corner of 41st and Alameda, you'd cross the street and walk into the farm. It was ten degrees cooler. And you had a situation where little kids in a heavy gang zone could run freely among the plants and feel safe."

Tezozomoc then took the stage, and in front of the eighty-or-so people squinting in the now baking sun, he announced that the struggle for the land at 41st and Alameda was back on the agenda. Eighty-five acres of land in a rural farming community is nice, but being able to grow food right in the communities that need it most is still the ultimate goal.

My conversation with Quigley rang in my ears as I careened along Highway 101, back into the tangle of shopping malls, twenty-four-hour gyms, and fast-food outlets that dominate the streetscape of North Los Angeles's San Fernando Valley, once a great wheat-growing area. "At some point, this industrial tack that we've been on, we're going to fondly remember the end point of that and realize 'you can't eat that.' And that little green square, as seen from the air, to me is the first space, the first move. As soon as we can get that back to green, that's going to have a huge ripple effect. When we get it back green, it's going to accelerate everything."[13]

PAVING EDEN

These are just two stories about the deeper issues at play with land use in the United States. The prevailing rule is that private ownership is still more important than public good, and this governs what type of ownership and use policy is allowable and what is not.

One of the many ironies of our human population explosion is that we are paving over the land that feeds us. Cities have, for the most part, sprung up on or near the most fertile, productive food-growing land. According to the American Farmland Trust, 91 percent of fruit and 78 percent of vegetables grown in the United States are in urban-influenced areas. You cannot talk about agriculture anymore without talking about urbanism.

We mistakenly assume that we can simply move our food-growing away from our cities and onto other farmland that exists out there just beyond the city limits. At some point, we will have paved over and contaminated the outer edges of our region's viable farmland. One day, we may find that we have paved ourselves into a corner.

Chapter 9

VANCOUVER
Canada's Left Coast

**We've moved to cities and we think the
economy is what gives us our life, that if
the economy is strong we can afford
garbage collection and sewage disposal
and fresh food and water and electricity.**
—Dr. David Suzuki, cofounder,
David Suzuki Foundation,
scientist, broadcaster, writer, Order of Canada

In 2009, the City of Vancouver, British Columbia, issued a document
called "Vancouver 2020—A Bright Green Future."[1] Three full pages
out of the seventy-four-page report detail the city's commitments to supporting local food and urban-agriculture initiatives. According to the
report, there would be an edible landscaping policy that obliged all city
facilities to incorporate 25 percent edible landscaping, including green
roofs. There would be collaboration with other alternative food enterprises in the city—urban farming outfits, small-scale food producers, a
café, community food educators—to create a food hub as the heart of a
local food system. And the city would support urban farming through
favorable tax rates for property owners and developers who would convert unused land to urban growing space. Overall, Vancouver would aim
to reduce the carbon footprint of its food by 33 percent. To achieve its

Multicultural Vancouver is bursting with community gardens and public green space. This is the sign to the entrance of Vancouver's Strathcona Community Garden. Photo by author, January 27, 2011.

goal of being the greenest city in the world by 2020, Vancouver will have to out-green London, Sydney, Copenhagen, New York, Portland, Seattle, San Francisco, Chicago, Toronto, Berlin, Paris, and Stockholm—just a few other cities with this goal in mind.

Unlike in many cities, this is not just bluster from the public relations machine at city hall. Vancouver actually takes being a green city very seriously, and its efforts are having an impact. *Economist* magazine has ranked Vancouver, Canada, at the top of its World's Most Livable Cities for several years running. In 2011, Vancouver again took top spot, followed closely by two other Canadian cities that are discussed in this book, Toronto and Calgary.

CITY FARMER: MICHAEL LEVENSTON

The irony that I was not able to find Michael Levenston's demonstration garden and office because of all of the *other* community gardens that surround it was not lost on me as I wandered from garden to garden asking directions. I quite literally couldn't see Levenston's City Farmer's head office— "Canada's Ministry of Urban Agriculture"—for all the flowers and food.

Even major hotels have rooftop food gardens. The terrace herb garden and orchard at the Fairmont Waterfront Hotel in downtown Vancouver, British Columbia, is home to over sixty varieties of herbs, edible flowers, fruit trees, and vegetables, as well as the hotel's six honeybee hives. Photo by author, July 19, 2010.

Community garden plot in the Kitsilano neighborhood, near City Farmer's office, Vancouver, British Columbia. Photo by author, July 22, 2010.

In other cities, I had grown accustomed to creeping along streets looking for a square of green that would let me know that I had finally found the urban food garden I was looking for. In Vancouver, I often had to confirm that I was in the *right* community garden, on that block! A 2007 survey by the City of Vancouver showed that 50 percent of Vancouver households with personal yard space grew some food there.[2] The city's website lists fifty-five registered community gardens with 2,200 plots available to Vancouverites who don't have a yard. So it's not really much of an exaggeration to say that urban food growing in Vancouver is more the rule than the exception.

This comes as no surprise. Vancouver is one of the rare Canadian cities that has a year-round growing season. In terms of plant hardiness zones, which are used to determine where various types of trees, shrubs, and plants are likely survive, Vancouver is a Zone 8, based on its average climatic conditions, as are London, Seattle, Portland, and even Atlanta, Georgia. Vancouver also has a well-worn rebellious streak with founders' status in both social activism and environmentalism that oozes beyond the national border. Vancouver is the birthplace of Canada's environmental movement that took place in the 1970s, and Greenpeace was founded in Vancouver's Kitsilano neighborhood, where Levenston's world headquarters and famous demonstration garden is located.

With the help of a couple of gardeners puttering away in their community garden plot along an old railway right-of-way turned strip of gardens, I finally found City Farmer—not an official embassy-like building emblazoned with maple leaves but a cobalt-blue house on a tranquil residential street.

CANADA'S OFFICE OF URBAN AGRICULTURE

Levenston first tackled the idea of urban food production as a means of energy conservation as part of a self-directed summer job in his late twenties for the Vancouver Energy Conservation Centre in 1978. The summer job didn't last, but the idea of raising the awareness of the environmental benefits of urban food production stuck.

Growing food in the city was a radical idea in those days, even in the nonconformist Vancouver of the 1970s. It was even a counter-counter-culture move, as many of Levenston's contemporaries were leaving city life behind to go back to the land to set up organic farms. But Levenston knew rural farming wasn't for him. He'd spent two summers in his youth tree planting, a potentially lucrative but backbreaking seasonal job that entails dawn-to-dusk digging and planting of tree saplings in remote locations. The heat, the bugs, the isolation, and the extreme physical demands were unbearable. In short, he found the experience excruciatingly unpleasant.

"I was a city boy from Toronto. I wanted to go 'back to the land,' *in the city*!" Levenston laughed as we sat on the shaded porch between his narrow, galley-like office looking out over City Farmer's garden in full midsummer production.[3] Urban agriculture was Levenston's way to scratch his environmental, agricultural activism itch without having to give up the creature comforts of city life.

Levenston and a group of friends decided that they'd have to forge their own path. They started City Farmer as a nonprofit society and cheekily appointed it as Canada's Office of Urban Agriculture. "It was so outrageous at the time," Levenston admitted, "that we didn't even worry about it." (They had no affiliation or sanctioning from the Canadian government and probably weren't even on its radar.) City Farmer's newspaper soon followed, as Levenston had previously worked as a journalist for the student newspaper in his university days back in Ontario.

In 1981, Levenston talked his way into a piece of land from the city, rent-free, to start an urban-food-growing demonstration garden at the corner of Sixth Avenue and Maple Street. The former residential lot, near a Canadian Pacific Railway track, was so polluted that it took two years to remove the contaminated soil and replace it with viable organic compost and dirt.

He enlisted local gardeners and other volunteers as the plot took shape. The garden was open to the public, not just as proof that food could grow in the city on a residential lot, but so that people could come to the garden and learn directly from the more experienced gardeners. It was an outdoor schoolyard to instruct and inspire would-be city farmers and community gardeners to plant their own beds with everything from summer squash to winter kale.

Levenston was also consulted on the burgeoning community gardening movement that caught fire in Vancouver in the mid-1980s and early 1990s, when most municipalities in Canada and the United States were ridding themselves of derelict gardens that had fallen out of favor. In 1985, Levenston helped start the Strathcona Community Garden, a three-acre, two-hundred-plot community orchard and garden east of downtown, now a showpiece in the Vancouver urban "greenscape."

Always on the lookout for new ways to spread and share urban-agriculture information, Levenston recognized the potential of the Internet early on. On October 15, 1994, City Farmer's website went "live," and, not surprisingly, it was the only online source of information for urban agriculture. It attracted readers from all over the world.[4] Within a few years, the site had a global reach, getting traffic from over 150 countries.

Levenston now mainly curates urban-agriculture stories and headlines from around the globe on his site. It is still the global "go-to" site for news from the frontiers of urban agriculture. "I call it the CNN of urban-agriculture news," Levenston jokes, but he's not that far from the truth. People send Levenston links, and he curates the site, usually posting a few dispatches per day, in between the daily tasks of running the City Farmer demonstration garden and other hands-on tasks. City Farmer's demonstration garden reaches a local population, and the Internet helps it reach the world. "I guess that's why I work around the clock. They're both valuable."

The city pays Levenston to educate and direct the Compost Education Centre. (Making and using compost for urban agriculture from food and yard waste is part of the city's green vision.) And the 2,500-square-foot (230-square-meter) garden and small office, which is now worth a fortune among Vancouver's notoriously stratospheric real estate prices, is still rent-free from the city. The City Farmer gig has been his only job since 1978. The telephone rang nonstop with composting questions the mid-July day I was there. "Calls come from all over North America," said Levenston, but the majority come from Vancouver.

Levenston is proud of the fact that he's been able to make a career out of his passion for urban agriculture and that this has been his job for over three decades. He knows he got lucky with the land from the city, and he's managed to justify his full-time salary plus the equivalent of another full-time salary for his staff. It certainly hasn't hurt Vancouver's image as one of Canada's greenest cities, either. Levenston hosts journalists and film crews from places like South Korea and Germany. He even gets visits from Cuban delegations, the undisputed urban-agriculture heavy-weights. One Cuban woman proclaimed Levenston's garden Mecca, he told me. "I'm not Mecca," he countered. "This is just a modest garden!"

Levenston then dashed off to a corner of the garden near the new electric composter he's been experimenting with. He spent close to an hour with a local television news crew. The city has been grappling with its municipal waste problem. A proposed incinerator doesn't sit well with most Vancouverites, so Levenston was showing off this large-capacity composter, suggesting that it might be a good option for apartment blocks or neighborhoods.

I scribbled in my notebook, taking inventory of the diversity of the garden's produce as Levenston worked with the television crew. There was a gooseberry bush, taller than me, dripping with ruby-like translucent berries. Armfuls of silvery cabbage, red-tinged lettuces floated in a carpet of violet-colored lobelia, a common ornamental flower. Alpine strawberries crept along the dirt. Huge stands of mint, sage, and other culinary herbs crowded each other. Several cardboard boxes lined with garlic bulbs were drying on the wooden porch in the shade between the garden and Levenston's narrow galley office. And I soon stopped counting the gardeners and visitors who were coming and going and grew used to the compost hotline ringing continually in the background. The place was a strange mix of constant activity and calm in the middle of a major city.

As Levenston returned to the porch after his television appearance, I

Cold frames and raised beds in the City Farmer garden, Vancouver, British Columbia. Photo by author, July 19, 2010.

decided I'd better ask the person who saw the urban-agriculture revolution coming three decades ago if he knew how far it would go. Would we be sourcing most of our food from urban gardens in the foreseeable future? Would we have commercial urban farming enterprises in cities all over North America? What did he think of the spate of "future scenarios" of urban farms on rooftops or glass-and-steel thirty-story vertical farms?

Urban agriculture, as seen by Levenston, is part trend, part necessity. It's cyclical, and it comes back when people get nervous about food scares, or the economy faltering, or the cost of food, or an energy crisis. "Obviously, today, people need it."

Then he coolly added that for him, urban agriculture will always be about what individual households are doing in their back or front yards. Levenston was the first pro-urban farmer I had come across to put the movement into perspective. "You're going to take some veggies, soft fruits, and herbs. And you're going to either grow *this* much or *that* much. You're either going to have access to vacant land or not. But it's not going to be cows. It's not going to be rice fields. It's not going to be wheat." Levenston then admitted that he doesn't even track the produce that comes out of the demonstration garden.

He worries that urban densification will actually hamper people's ability to grow food on a household scale. "A garden works when you have enough room to actually do something and be productive. So there's a conflict within modern urban agriculture."

"I have a lot of worries about it being real," he continued, rather soberly. He pointed to the example of Vancouver's urban chicken bylaw change in the summer of 2010. The bylaw was put in place to track the number of urban chickens and their locations around the city. It could have been the proof that urban chicken activists needed that it was already happening in the city without any major downside. "Only seven people signed up to register their chickens with the city." His point was that a wave of new policies are symbolic of change, but did they really amount to change in itself?

But then his cautiousness gave way to optimism. "It's hard to knock the enthusiasm and the productivity that I see on the blogs out there. It's good as long as people are reading critically." He finally admitted, however, that the mood *is* different lately. "We don't know where our food system is heading. We don't know where our farming is headed. There seems to be support for the first time from municipal governments. The urbanization of the developing world is new and uncertain. We now have massive urban populations." The variables have changed, so maybe the food system will change.

YMCA INTERCULTURAL COMMUNITY GARDEN

Levenston is a connector. After a morning spent at the City Farmer head office, I had appointments all over Vancouver with key people in the

The YMCA Intercultural Community Garden (now Downtown Intercultural Gardeners Society garden—DIGS), Vancouver, British Columbia. Photo by author, July 20, 2010.

city's urban-agriculture movement. The very next morning, in fact, I was standing on the rooftop of a century-old hospital in downtown Vancouver talking with David Tracey, author, arborist, environmental designer, guerilla gardener, coordinator of the Vancouver Community Agriculture Network, and coordinator for the YMCA Intercultural Community Gardens Project at St. Paul's Hospital.

"It's unusual in community gardens, as it's not exactly about the food and the ecology, environment, nutrition, and all those other things we have community gardens for," Tracey explained as we stepped out of the pastel-colored hospital corridors and into the July sunshine.[5] "This is one attempt to make Vancouver a more welcoming place."

Vancouver is one of Canada's most multicultural cities. This is part of its appeal to newcomers and Canadians alike. While "everybody mostly gets along in Vancouver," he continued, subtler types of racism obviously exist. There are not always intercultural exchanges among the different immigrant groups, even though they share so much in common. "So they get along, but they don't really do anything together. This project is an attempt to see if we can reshape Vancouver into a place that is actually *inter*cultural, where different cultures really are working together, and doing things together."

Tracey's first job was finding a suitable garden space in downtown Vancouver. He lucked out. As it happened there was an enormous outdoor space, renovated in the 1970s, on an upper terrace at St. Paul's Hospital. There were already white window-box-style planters attached just below the perimeter railing where the terrace overlooked the skyline and glimpses of ocean beyond. In the middle, raised concrete planters with trees, bushes, and large grasses had been placed throughout the 21,500-square-foot (2,000-square-meter) area. But the boxes were at waist height, making them accessible to those in wheelchairs and to seniors; they were just a comfortable height to work at. Tracey secured a deal with the hospital administration to establish a community garden. Tracey acknowledged that it was a big step and quite unprecedented for a hospital to invite a group like this to use space for food growing. That said, the community gardeners there have also helped the hospital with a couple of problems.

First, the space was being used inappropriately by drug dealers, as it was sort of a public space, and the park-like setting gave them a bit of privacy to conduct business. "Since we've been here, it's been a little less comfortable for the drug dealers. If there are gardeners hanging around maybe with kids, it's just harder to come and deal."

The other service the gardeners provided was a landscaping service of sorts. One of the first activities the gardeners did was to remove forty yards—meaning 1,080 cubic feet, or 30 cubic meters—of old, spent soil from the planter boxes and then bring the same amount of high-quality soil up to refill them. "We had to bring this *through* a hospital! We started in the loading bay, rolled past autopsy, went into the elevators, and unloaded the soil up here." Thankfully, the hospital was getting ready to dispose of some wheeled plastic laundry bins. "So we recycled their laundry bins for them." They became giant dirt-filled wheelbarrows. Tracey pointed out a giant plastic laundry bin in the makeshift "garden shed."

"Often when a new community garden starts up, a lot of people will turn out, but they tend to be established Canadians who understand how the system works. Often they tend to be white, middle class." Language barriers and social inhibitions prevent immigrants from asserting their right to community garden space. This garden was going to be different.

According to Tracey, the "downtown peninsula," the 1.5-square-mile (3.8-square-kilometer) collection of four inner-city neighborhoods where the community garden is located, is made up of 40 percent non-Canadian-born Vancouver residents. Yet statistics show that it can take up to ten years before immigrants to Canada feel suitably at home in their new city and able to fully participate socially and economically. Even then, as Tracey mentioned, there's often a lack of intercultural interaction.

This is why 40 percent of the spots in the garden are reserved for non-Canadian-born participants, to reflect the demographics of the surrounding area and foster some interaction between the various cultural groups that call the area home. The gardening participant guides were printed in Mandarin, Spanish, Russian, Farsi, and English. The garden's rules also required each gardener to complete training and education in intercultural communications, diversity, antiracism, and antihomophobia, as Vancouver's downtown peninsula has a large gay population.

In 2010, the community garden's inaugural year, fifty-three gardeners joined. Tracey admits that much of the first year was taken up with organizational and start-up activities. Part of the training process was also to educate the gardeners on how to become part of a self-governing body. Tracey needed to train them in how to set up their own nonprofit, how to establish bylaws, how to deal with gardeners who

Davie Village Community Garden. The terrace at St. Paul's Hospital, which serves as the home of the Downtown Intercultural Community Garden/DIGS, is visible to the right of the Davie Village Community Garden sign. Photo by author, July 20, 2010.

don't cooperate or work outside the bylaws. And then there was organic gardening training for many who had never sowed a seed in their life before this.

Each gardener was assigned two planter boxes, one in full sun, the other in partial shade. Each rectangular box holds just over seven square feet of soil, so each gardener has about fifteen square feet of space. The irrigation system was already part of the planters, though Tracey admitted that they were still working out the idiosyncrasies with the pre-existing irrigation.

This meant that the gardens got a late start. Nevertheless, by mid-July, ripe tomatoes were hanging on plants, peppers were filling out, and a few ambitious blueberry and raspberry canes were already producing fruit.

There are hundreds of other boxes available for expansion if the gardeners wish, though some areas of the rooftop haven't seen sunlight for a hundred years and might never be viable growing spaces. Tracey's contract with this project was for only a year, so in May 2011, the group "graduated" to their own autonomous, self-organizing, self-governing, self-supporting community garden, and renamed themselves the Downtown Intercultural Gardeners Society (DIGS).

DOWNTOWN EASTSIDE VANCOUVER'S SOLEFOOD FARM

In 2010, a 17,000-square-foot (1,580-square-meter) farm sprouted from a derelict parking lot next to a run-down hotel in a notoriously rough part of downtown Vancouver. I decided to drive there, rather than walk the twenty minutes in the morning July sunshine, a decision that seemed silly on the surface. But this is Vancouver's reality. Just blocks from gleaming office towers and luxury hotels, countless Starbucks locations, and million-dollar condos, there are blocks where extreme poverty, mental health issues, the sex trade, and drug deals can be seen in plain daylight, and they have changed the landscape entirely. It's unnerving as a visitor. I had visions of taking the wrong route and finding myself in a dangerous situation. Vancouver's Downtown Eastside is not for the skittish.

I won't dwell on the well-known social problems in this strip that stand out so boldly against Vancouver's overall image as a health-obsessed, affluent, cosmopolitan city. Moreover, I'm not qualified to judge why it is the way it is. I just know that "poverty tourism," a strange type of voyeuristic tourism of impoverished areas of Vancouver, among other places, has become a source of annoyance to those whose home is the Downtown Eastside. My interest in the farm, though, happened to include the fact that it is located within a neighborhood that isn't otherwise associated with high-value market vegetables.

When I arrived at SOLEfood Farm, I circled the perimeter chain-link fence until I finally got the attention of one of the farmers. It took a few minutes for someone to find the key to the padlock to let me inside the farm. But as I stepped through the gate, the familiar smell of plant life hit me, and I realized how nervous I had been while waiting to be let in. Yet, once inside, it was a different world. A pigeon was taking a rambunctious bath in puddles that were left over from the morning watering. The acidic smell of wet pavement mingled with the sweet, rich smell of wet earth and photosynthesis.

Seann Dory, the farm's manager, had been at work since five o'clock that morning—Wednesday afternoon was one of four market days a week—and he was keen to tell the story of SOLEfood, so far really only a locally known phenomenon. Dory, with a few days of facial scruff and a wool flat cap that made him look more like an independent musician than a farmer, technically had been a farmer for only a few months.[6]

As the manager of sustainability at United We Can, a grassroots inner-city community nonprofit group that fosters and runs environmental social enterprise initiatives, Dory was searching for a new environmentally friendly business venture when the idea of starting an urban farm was raised. Stakeholder groups under the United We Can umbrella were meeting to brainstorm enterprises that would create jobs in the inner city through green initiatives. "Urban agriculture seemed to have the most energy behind it, and the most support at a city level as well." But the year-long search for a suitable site in the Downtown Eastside proved fruitless despite the generous tax incentives the city was offering to businesses that converted unused land or brown space into green space.

Ultimately, the Astoria Hotel, a shabby rooming house on a major thoroughfare through Vancouver's most desperate neighborhood, offered up its rubbish-strewn, rat-infested parking lot that had become a dumping ground on the block. The problem was that the offer came in from the Astoria Hotel just days before the deadline for the municipal tax evaluation. Regardless, Dory and fifty volunteers literally built the farm on October 31, 2009, the day before the cutoff to qualify for that year's tax break.

With a de facto urban farm on Dory's hands but no actual gardening experience or knowledge under his belt, the winter of 2009/2010 was a

SOLEfood Farm's yard, Vancouver, British Columbia.
Photo by author, July 21, 2010.

flurry of consultations. "None of us had any farming experience, so we surrounded ourselves with experts. I grabbed as many master gardeners as I could find, and I put them on our advisory board. We met quite frequently for a while." One of those farmers happened to be Michael Ableman, who now farms on Salt Spring Island, on British Columbia's coast. Despite running his own 120-acre (48.5-hectare) Foxglove Farm, Ableman joined the farm's board of advisers and gave Dory and the farm staff invaluable hands-on mentorship throughout the first year.

"We're making mistakes regularly," laughed Dory, yet I was becoming increasingly more distracted by a woman nearby washing the most beautiful white-tipped red French breakfast radishes I'd ever seen.

Not nearly as impressed by the morning's radish harvest as I was, Dory pointed to the raised bed growing glossy green peppers. "Those stakes are inadequate. We have to redo those." Then he rattled off a list of other missteps: watering issues, overharvesting issues, bad seeding technique. "But we're still able to grow here."

Besides the Technicolor® breakfast radishes being washed and bundled for market and restaurant deliveries, gardeners were picking ripe, perfect tomatoes from the plastic-draped "high tunnels," essentially makeshift greenhouses with curved plastic piping supporting heavy clear plastic, just tall enough to walk through. And they were on their third cutting of salad greens, a mix of red Lolla Rossa, green oak leaf, and arugula baby lettuces. Another young farmer, Jordon Cochrane, happily posed for me as he harvested dark-green "lacinato" kale leaves, sometimes referred to as dinosaur kale for its alligator-skin-like texture. The farm was also growing collard greens, garlic, spinach, basil, eight types of tomatoes, seven types of sweet peppers, and eight types of hot peppers.

"Want to see our pinpoint seeder?" Dory asked with that I've-got-a-new-toy inflection, pointing to a hand-sized piece of farm equipment with outer wheels, a handle, and five mini-hoppers that release seeds at perfect intervals in perfectly straight rows. He also noted that it didn't take long for bees to find the farm. He has even noticed some "volun-

(*left to right*) Seann Dory (manager) and urban farmer Jordan Cochrane harvest dinosaur kale at SOLEfood Farm, Vancouver, British Columbia. Photo by author, July 21, 2010.

teer" seedlings growing in cracks in the pavement. And the recent discovery of worms in the garden beds prompted him to shout out, "Worms! We've got worms!"

I suggested that these premium market veggies look like they are out of the price range for the people shuffling past the farm. Dory bluntly agreed but quickly pointed out that SOLEfood Farm's goal was not to feed the neighborhood. Its goal was to provide real-life skills training for those who go through the training so they can find reliable work and support themselves. And to do that, the farm needed to be a commercially viable business, and for that, they needed cash flow. This meant that they were selling about 90 percent of the produce directly to high-end restaurants and the farmers' market crowd in upscale areas. The remaining 10 percent would go to the surrounding community organizations working on food-security issues. Staying true to United We Can's social enterprise model, providing subsidized fresh food is not the answer to the Downtown Eastside's problems, but providing skills training that leads to an income stream—as an urban farm provides—offers long-term sustainable security.

That said, the immediate benefits to the community were evident as I sat there watching the activity both at the farm and around the chain-linked perimeter. The plants, bees, and birds were clearly a tonic for area residents. One man, Mario, pressed his face into the chain-link as I was talking to Dory and asked about "his beans." Dory explained that Mario was an enthusiastic volunteer who often came around to check on the farm. Even behind the fence and padlocks, SOLEfood Farms was a little Eden for the area residents to focus on.

By the end of the 2010 growing season, SOLEfood Farms had grown ten thousand pounds of high-market-value produce. It was enough to pay the wages of two full-time and five part-time staff. As a start-up, farming or otherwise, it was a raging success. Dory hopes that other commercial urban farms take inspiration. "In one year, we've grown one of the most successful social enterprises in North America and it's in an urban setting."

Vancouver's New City Market

As I poked around in urban agriculture, food policy, and community-development circles in North America, Herb Barbolet's name kept bubbling to the surface. After a few e-mail exchanges, he agreed to chat with me about his involvement in what is perhaps his most ambitious project in his forty-year involvement in local agriculture, community activism, food-policy creation, and sustainable food systems. For the past few years, Barbolet—who brings some serious academic credentials to the table with a master's in community development and two PhD programs (ABT) in community development and in community planning and political economy—has been a member of a collaborative called Local Food First. Its mission is none other than to reinvent Vancouver's food system around local supply and distribution. If it succeeds, Vancouver could make major strides toward breaking the chain that the global industrial-supermarket model dominates even in one of the most leading-edge sustainable cities.

In his twenties, Barbolet lived in an "intentional community" in Vancouver while farming organically just outside of the city. Exotic organic salad greens in the 1970s were novel and somewhat of a tough sell, but for a decade he managed to sell high-value greens to health- and taste-conscious Vancouverites. This would be just the beginning of his role as bridge between the rural food producer and the urban eater.

In 1993, Barbolet founded FarmFolk/CityFolk, to raise awareness of food growing and farming with consumers in the city. (Farm-Folk/CityFolk is still an energetic education and advocacy group for local, sustainable food relationships between producers and city dwellers in British Columbia.) That same year, he helped form the Vancouver Food Task Group, a grassroots group that was beginning to look at the complicated relationships between food security, nutrition, agricultural land use, and civic planning, among other broad areas. By 2003, Vancouver's city council passed a motion in support of the development of a "just and sustainable food system for the City of Vancouver." A sustainable food system was defined as one in which "food production, processing, distribution and consumption are integrated to enhance the environmental, economic, social and nutritional health of a particular place."[7] The motion also supported the forming of the Food Policy Task Force, which in turn led to the creation of Vancouver's Food Policy Council in 2003 to advise the city council on food-policy initiatives at the city level and later to the adoption of the Vancouver Food Charter in 2007, which gave municipal endorsement to the vision of a city food system that benefits the community and the environment it serves.

Local Food First began to discuss scenarios for an alternative, local food system in Vancouver in 2006. By 2010, the group revealed its preliminary business plan and design drawings for the New City Market. The design included plans for a processing facility for value-added food preparation and production of local foods, a permanent and year-round market space with storage, wholesale, and retail capacity for local farmers and producers, and a space for community education and outreach and marketing. And it would be a self-sustaining business entity, not subject to the whims of city council budgets. Vancouver city officials are currently very much behind this proposal, as stated in the city's 2020 goal-setting document, which declared that the city wants to create a food hub model that, if successful and adopted, would change the supply lines for the city's foodshed dramatically. If the stakeholders in the New City Market can actually make this leap from the page into the community, it might be a new civic model for urban food systems elsewhere.

"Those of us who have been working in the alternative food system for decades have been working with 1 to 2 percent of the food system," Barbolet explained as we met over coffee at Vancouver's famous Granville Island Market. "When Walmart decided to go organic, they did more in one year across the world than we had done in decades."[8] This was just further proof for Barbolet that the alternative food system needed to organize, to consolidate its buying and selling power, and to scale up to the point where it could be a true challenger to the industrial food/supermarket model.

With so much change happening so rapidly in the mainstream food system, Barbolet and his colleagues saw the opportunity to get in on that change and started to think about how to support small producers, protect local farmland, and service the demand for local food. At Local Food First, they mused about how to connect the twenty thousand smaller and mixed-farming outfits that could supply the metropolitan Vancouver area with eager restaurant owners, retailers, and households clamoring for more access to local food.

The result of those musings was a design document and business plan for a 30,000-square-foot (2,780-square-meter) vertically integrated food hub in the middle of Vancouver. The proposal included space for wholesale and retail buying and selling, an indoor-outdoor year-round farmers' market, space for food storage, distribution, processing, education, and research and development facilities. Cafés and restaurants would round out the business mix.

When Local Food First unveiled this proposed food hub in late 2010, it was already the result of three years of work, and the group was "three to five years" away from breaking ground, or not breaking ground, as Barbolet put it. The ideal scenario was to retrofit existing buildings when possible. In fact, Local Food First identified the city-owned buildings that now stand where Vancouver's original city market–wholesale café was first established in 1908.

Barbolet acknowledged that the New City Market, currently budgeted at $8 million (Canadian), is uncharted territory both as an urban redevelopment scenario and as an alternative food system. But having worked with the city planners and on the development of a nuts-and-bolts feasibility and business plan, the new food hub might soon be a reality. "The fact that on a municipal level the city government is 'getting it' changes everything. It's script-changing." Overall, the momentum has become undeniably positive, and other cities are looking at the concept. "Everyone sees this as an opportunity to reinvent the city around food."

The problem is that few models even exist. Barbolet pointed to The Stop Community Food Centre in Toronto, Ontario, which began as a food bank in 1982 and still provides front-line food services such as a food bank. It is now a successful community food center with a drop-in soup kitchen, community food and gardening education center, farmers' market, and so much more. The Stop, according to Barbolet, wisely got away from the charity model and moved toward the social enterprise model. "What we've learned [about an alternative food system model] is that it has to be closer to the standard business model." To move this plan forward, Barbolet believes that various food-security initiatives in the city have to be convinced to concentrate their capital on establishing this new infrastructure. But funds for the capital investment needed to build the New City Market could come from the general public, which, Barbolet likes to remind people, is "still where the vast amount of funds are."

"Like an Initial Public Offering," I casually joked.

"Exactly," he smiled back.

Barbolet, like Levenston (of Vancouver's City Farmer urban-agriculture demonstration garden), was cautiously optimistic about the future of food in his city. He acknowledged that Vancouver was riding another wave of enthusiasm for urban agriculture and local foods, but he wondered out loud what portion of that would amount to lasting change. Despite the momentum and good intentions at work in key cities at the forefront of urban agriculture, Barbolet feels that so much more has to change, and sometimes, urban agriculture can even serve as a distraction or as only a temporary solution. Barbolet then laid out two extreme scenarios of urban agriculture. One was a dystopian view of gated gardens "where people with submachine guns protect their cucumbers." The other was an idyllic description of his travels through the streets of Ho Chi Minh City, a place Barbolet recalled where food grows in the streets, or Bologna, Italy, where everybody understands why food is important.

As Barbolet explained it, "the world has always fed itself," adding, though, that there's always been starvation. "The fact that Ethiopia, in the 1980s at the height of the worst of the famine, was still exporting food, and what the world did, in response, was to give them foods. It destroyed local farming. What the world *didn't do* was to provide infrastructure so that they could get trains across the mountains to distribute the food. It's the controls in the system that make some people rich and poor people even poorer. Until *that* is confronted, we can't confront any of it."

Barbolet intimated that those levels of control might just be subverted at the local level, which is undoubtedly what excites him so much about the idea of alternative food systems like what Local Food First is proposing with the New City Market. "You'll see the reemergence of the city-state," he proposed just as we left off, and I imagined the city-state of Vancouver in the not-so-distant future in charge of feeding itself.

SPINNING THE URBAN FARMING MODEL

Kelowna is 250 miles (400 kilometers) east and slightly north of Vancouver. It's in a dry, desert-like valley in the mountainous British Columbian interior where historic fruit orchards are now becoming mostly grapevines for the expanding Canadian wine industry. It's also the lakeside playground for wealthy Vancouverites from the West Coast, and for oil-patch-rich Albertans from the east. Kelowna has only around 110,000 residents, but it has sprawled out with big-box stores and traffic problems as it has grown. In short, it's not the kind of place I expected to find a vegetarian, politically charged, bicycle-powered indie-rock musician turned urban farmer. But then I met Curtis Stone.[9]

We stood in between rows of sharp-edged mizuna, also called Japanese mustard greens, the color and luminosity of lime-flavored Jell-O®, and rows of mustard greens with deep-purple leaves and emerald stems in what was rather recently someone's front lawn. Stone picked a purple leaf and asked me to tell him what taste it had. After mere seconds of chewing, there it was: the unmistakable peppery, vinegary tang of Dijon mustard.

There were rows of reddish oak leaf lettuce, spinach, sweet basil, Thai basil, carrots, beets, radishes, flat-leaf parsley, and so on. The only unruly part was a patch of flowers growing in the shade of a hundred-year-old European beech elm tree on the corner of the lot. Hundreds of Spanish onions and red onions were drying in a makeshift shed at the junction of three patches of garden on a corner lot where a house stood for several decades before it burned down and had to be demolished.

"It was a weedy hole," said Stone about this space, now ablaze with rows from dark reds to the lightest of greens. The lot was less than shovel-ready when Stone struck a deal with the owners—a family he'd grown up knowing—that he would clean up the site, fill the hole where the house once stood, and plant an urban farm on top. Stone promised to look after the property and give the family free produce in return for rent-free land. Both sides agreed that it was a win-win situation.

Aside from the hole and the weeds, Stone removed "over three hundred used needles" and other unsavory debris. "The first thing I did was build this fence," he said, motioning to the five-foot-high wood fence around the perimeter that protects this corner lot turned row crop oasis from the busy thoroughfare of the main roadway in town.

This was just the primary lot that Stone had planted as part of his

Urban farmer Curtis Stone of Green City Acres, Kelowna, British Columbia, at one of his garden plots. Photo by author, September 12, 2010.

patchwork urban farming enterprise. He had a strip of land with two hundred tomatoes on 6-foot (1.8-meter) trellises just around the corner. Another nearby plot was planted with salad greens. In total, he was juggling ten sites—all donated—but he said that he'd already started scaling down. Ten was too many, and the highest-maintenance crops, salad greens, were also the highest-value crops. He was going to pay more attention to those plots and plant slower, more independent crops on the plots farther away. He was a one-man show, and he got around only by foot and bicycle, so every efficiency he could coax out of his day, he did.

"I'm super methodical," Stone admitted as I remarked on his charts and lists in his work shed—even the way his gardens were organized. "I just go hard at things."

Stone became a vegetarian at the age of sixteen. He then pursued a decade-long music career throughout his twenties by moving to Canada's epicenter of independent music in Montreal. He played guitar, bass, and keyboards with People for Audio, which he described as "post-rock" music. The band was moderately successful on a local scale and released a few CDs, the last of which was produced by Stone.

"I'd always had an interest in sustainability; I just never did anything about it. And suddenly, I just got so sick of being one of those people— they'd hang out with musicians and artists, all educated intellectuals with great ideas about everything, but nobody actually did anything. I didn't want to be that guy anymore."

Stone gave everything away except some music equipment and left for a bicycle trip down the West Coast from Vancouver to San Diego. Along the way, he checked out various organic farms. Riding a bike for 2,500 miles (4,000 kilometers), Stone told me, wasn't actually very difficult. And he didn't really learn that much about farming on his trip. But he really reacted to how people perceived his traveling so far by bicycle. "People are so inspired by you!" By the time he got down to San Diego, he knew that he wanted to get into sustainable food production. Stone realized that he could make a big impact by setting the example. He'd be the guy who walked the walk with sustainable food production.

He was interested in agriculture as a sustainable career and even talked to his dad, who had similar leanings, about getting a piece of land to farm. "This comes out of my disdain for hyper-consumerism and the way we live in the modern world," he explained. He noted that growing food was one of the few money-earning careers that would have sat well with him ideologically. But land, even in remote places in British Columbia, was too expensive. Stone, a self-described people-person and extrovert, was also single, and he realized that his chances for a social life, let alone a romantic life, would be nonexistent in the wilderness. "I had to ask myself, 'Am I really ready to go off and live in the bush with my *dad*?'"

He heard about urban farmers making "six-figure salaries on an acre of land" based on techniques, crop selections, and business plans specifically designed for multilocation, multi-crop urban market produce farming on anything between a half-acre to a full acre. This type of farming is called SPIN farming; *SPIN* being an acronym for "Small Plot Intensive," and it was pioneered by urban farmer Wally Satzewich in Saskatoon, Saskatchewan, in the early 2000s. After detailing the business plans on how to farm successfully in the city on a patchwork of residential pieces of land, Satzewich began training other would-be urban farmers in his SPIN methodology. With a decade of refining the process for both Canadian and US urban farmers, Satzewich and his business partner Roxanne Christensen calculate that a SPIN farmer can gross anywhere from $27,000 to $72,000 on just a half-acre (two thousand square meters) of total land in the city, depending on the types of crops the farmer grows.

The SPIN farming website says that it's a method that "removes the two big barriers to entry—land and capital."[10] SPIN farming takes advantage of the fact that many people in cities are willing to donate or rent part of their residential land for a minimal cost. If a SPIN farmer rents, the going rate is listed at $100 per thousand square feet (ninety square meters). And no large equipment is needed. A personal vehicle serves as the "farm vehicle," if needed. Shovels, rakes, and a hose are really the only requirements for tools. That said, a rototiller, likely the biggest expense if you want to buy your own, makes preparing the beds much easier.

SPIN farming relies on the labor of the farmer, on smart crop selection, on efficient growing techniques, and on direct access to customers via farmers' markets and other direct sales as its core model. By keeping costs low and efficiency high, a SPIN farmer, says the site, can gross over $50,000 per year on a half-acre (two thousand square meters) of land in a city.

Stone was skeptical, but he tracked down a few SPIN farmers and mined them for information. "Some didn't make a living at it, but some did. I felt at least confident that I could make a statement by doing this."

He drew up a business plan and brokered deals for land. (Offers of other sites came fast and furious at Stone as word spread about what he was doing.) He even started a composting program by giving "little white buckets to friends and a few restaurants" that winter. And then he bought a rototiller for one thousand dollars and a walk-in cooler for the same amount, staying well under budget.

Even before he planted his first row, Stone became a local media sensation. A few stories ran on his planned urban farming idea, and he "was an urban farmer six months before I was actually an urban farmer." When March came around, Stone was actually quite petrified that the whole thing would be a complete disaster.

It wasn't. His homework paid off, and the learning curve was steep. He picked up time-saving techniques as much out of necessity as comfort, like "flame weeding." Stone explained how much faster and easier it is to run a blowtorch along a row of seeds just before germination to incinerate any weeds, leaving rows clear for the new shoots that soon appear. "Much better than spending six hours on hand and knees weeding two hundred feet of carrots!"

He bounced back quickly from rookie mistakes. Restaurants jumped at the idea of ultra-fresh local produce, which he'd deliver via his custom-built bicycle trailer that he'd simultaneously use to pick up kitchen compost from those same restaurant clients.

It didn't matter that his salad greens were three dollars more per pound than what a chef could order through the industrial suppliers. Stone's greens were harvested the day of the order. They lasted for two weeks in a cooler. And they had more volume and body—so much volume for pound that chefs actually used less. "Their bottom line improved" by paying more, Stone explained.

The pent-up demand for high-quality local produce at the city's farmers' markets also worked in Stone's favor, because he could make as much as $1,500 per market day. In fact, market days, which can be grueling for any overworked farmer, were the highlights of his week. "It's just appreciation, all day." He drew energy from the interaction and the face-to-face connection with his customers. "I get to run my mouth about all the bullshit that still needs to change. And I have an open audience for it. And I make money at the same time. How good is that?" Green City Acres, his urban farming business, was booming.[11]

The eighty-hour workweeks, on the other hand, were punishing. "Which is not sustainable for a human, by the way," Stone advised. "Maybe if I had a partner there, feeding me all the time." Some days, he'd work for sixteen hours and then be too exhausted to cook for himself. "I'd be too tired to eat!" Stone also found the long workdays could be isolating. "It's such hard work. I've had breakdowns. You feel alone. You feel underappreciated. Kelowna is an easy place to feel underappreciated." (Just a month after I first interviewed Stone, he was awarded "Gardener of the Year" in 2010 by a Kelowna community beautification initiative called Communities in Bloom.)

I caught up with Stone again at the end of his first year, curious to see if his enthusiasm for his new career was still there, and, perhaps more importantly, if the model worked financially. Remarkably, in his first year of urban farming, Stone managed to grow two dozen different crops on 25,000 square feet, just over a half-acre, that is, of piecemeal city land. Despite the $7,000 (Canadian) initial capital start-up costs, Stone still came out ahead, grossing $20,000 (Canadian), rookie mistakes and all, for a half-year's work. Already, he was tweaking his product list—adding microgreens and focusing more on the salad greens that fetched the best profit—for the next growing season. And he was taking orders ahead of time for a CSA operation so that he knew he had a stable roster of weekly customers, which allowed him to have a more balanced workweek as well. "I won't have to do the eighteen-hour days anymore," he figured, because he could cut out one of the weekday farmers' markets now that he had CSA customers. And then he told me that he had set his income goal for his second year at $50,000.

Making a profit and setting such lofty goals has made Stone a bit of a controversial figure in the sustainable food community, especially in Vancouver. "There's an unspoken rule that urban farming and making money are not done," Stone quipped. But he thinks that commercial sustainability will be the lasting catalyst for change. And he has found an off-season tie-in. In the winter of 2010–11, Stone gave over twenty SPIN farming workshops and presentations on urban farming throughout North America.

In 2010, Stone gave away about fifty pounds of food each week, at the height of the season. "That's inherently anti-capitalist, the idea of sharing," he said gleefully. This, Stone explained, is part of his plan, to get people "comfortable with having gardens and with sharing." Because, as Stone puts it, if the system were to collapse, those with a capitalist mentality would approach the situation as every person for himself. And life would get awfully precarious awfully fast. But Stone wants to "go on, have a family, and do all those things that humans do."

Stone's ultimate goal, he told me, will be to show that a sustainable living outside the conventional grocery and industrial food system is possible in a city. He also knows that he needs to succeed in order to inspire other people to follow his lead. "Even if it's just inspiring them to tear up their lawns and start growing food just for themselves. Because if we can produce all that we need around here, then what do we need the food system for?"

Stone still gets riled up easily about people's lack of willingness to change their lifestyle. This grates on him. "People are into sustainability at an entertainment level. They encourage sustainable enterprises and initiatives, but I'm doing it. I'm pulling a four-hundred-pound rototiller around on my bike!"

"We need to get radical," he pleaded, as he said farewell; he needed to get back to his pedal-powered mobile urban farming business.

The Roots of SPIN

The "food not lawns" movement, and the 2006 eco-sensation book by the same name, already chipped away at the idea that residential yards had to be a fertilized, overwatered, unproductive lawn, bordered with wood chips and evergreens. But this movement was part political act— in many municipalities, there were laws against food growing in front yards, though most cities have repealed these laws in the face of changing social norms and common sense—and part pulling a few veggies from your front yard.

SPIN farming, on the other hand, was designed not as an ideology but as a franchise-like system, an entrepreneurial business model aimed at making money for the owner-operator.[12] And it cleverly gets around the two biggest barriers that first-generation farmers face: land and capital.

The system was designed by Canadian farmer Wally Satzewich as a new solution to a growing crisis in the farming community in the 1990s. It was a new approach to making a go of it as an independent family farmer. In the past few years, it has caught on like wildfire. There are now over six hundred SPIN farmers operating around the world, from Canada and the United States to Ireland, the United Kingdom, and Australia.

Between 1993 and 1999, Wally Satzewich and Gail Vandersteen lived in Saskatoon, Saskatchewan, but they owned a quarter-section (160 acres/65 hectares) of land forty minutes away. They grew organic market produce on just about ten acres (four hectares), and they'd sell this organic produce at farmers' markets back in the city. The rest of the arable land on the farm was leased to neighboring farmers for additional cropland. Like most farming families, there was off-farm income that kept them afloat. Vandersteen had a good job in the city at the University of Saskatchewan.

I called Satzewich on a mid-January morning in 2011, when winds were sweeping snow across Saskatoon at sixteen miles per hour (twenty-seven kilometers per hour), and the outside temperature was a seasonal minus-four Fahrenheit (minus-twenty Celsius).[13] We spent the first several minutes discussing weather, a Canadian Prairie preoccupation, because that year the snowfall had been heavy. Fifty inches (127 centimeters) had fallen, and the winter wasn't even half over yet. Even Canadians know that Saskatchewan's winter weather is not for the thin-skinned. Soon enough, however, we got around to SPIN farming.

As Satzewich tells it, organic certifiers used to come to the farm and continually harp at him about scaling up. "They'd say, 'Listen Wally, you've got to break away from the farmers' markets and start thinking about semitrailer-loads of organic potatoes being shipped to Vancouver.' That was the kind of model they were trying to hammer home to me."

There were other problems with the way Satzewich was farming. He was always worried about finding someone to lease the other hundred acres of cultivatable land year to year. Then there was frequent crop damage from marauding deer, unexpected late and early frosts, and a very short but intense growing season in the northern latitude. And the relatively small size of his operation didn't allow for as many efficiencies as a large farm can take advantage of. So, in the conventional way of thinking, it was "go big or go home."

Interestingly, Satzewich went home. He had always kept a garden plot in his yard in the city where he'd experiment with new crops. Eventually, due to constant deer raiding in the countryside, he started growing his salad greens in the city. He noticed that he could get his beds going earlier in the season in the city just from the small bump in heat. There were very few pest problems, and the greens grew happily with a longer growing season. He could get three crops off his city plot for the one he could manage in the countryside.

The tipping point came, said Satzewich, when a neighbor made an offer on his family farm. "By that point, I had a feel for what I could grow in the city." Worst-case scenario, he figured, he could grow salad greens part-time while Vandersteen kept her job at the university. They sold their farm in 2000.

That spring, Satzewich asked family and friends if he could convert some of their yards to urban farmland. They agreed. He also put an ad in the local newspaper, which got him a few more yards to work with. Over the next few years, Satzewich continued to refine his cropping techniques and timing, to select optimal crops for the climate and market sales, and to figure out how much land he needed and could comfortably work to make a true commercial go of it.

Satzewich then cataloged his learning curve in a monthly newsletter he wrote called *Market Gardening Concepts*. This small-plot, intensive urban-agriculture system became SPIN farming. "I just tried to quantify what I had learned and [hoped] that maybe it would help someone avoid the mistakes that I made." He also thought that if he could get across the idea that people don't need twenty acres to be a vegetable farmer, they could start to think about an acre of land instead, and what they could do with that. "For me, it was defining farming at a much smaller scale of production. And the main thing was to keep [sub-acre farming] as a choice." This was an epiphany even for Satzewich. "I always thought that if you were at an acre, you wanted to get away from that as soon as possible."

But the economics of scale-up just didn't make sense. As we spoke about the mathematics of even medium-scale farming, anyone with only a cursory knowledge of agriculture could see the problems. Wheat prices in North America are still quoted in price per bushel, and you get about thirty bushels per acre, Satzewich explained. Even on the upper end of the wild swings in wheat prices—say, $10 per bushel—that's only $3,000 of gross revenue per acre. "Then [a farmer] needs a million-dollar tractor. And they have to take out a $100,000 bank loan just for their input costs," he continued, referring to the cost of seed, fertilizer, pesticide, herbicide, crop insurance, and fuel for the tractor. That is all up-front cost. "Given the volatility of our growing season, a lot of farmers I know just aren't willing to put those costs up. It's just not worth it." So Satzewich continued to build his urban farming model, figuring out how to make the most money on a small enough land base so that he and Vandersteen could do the bulk of the work themselves with the lowest overhead.

El Niño weather patterns dominated in 2010. It was a difficult year because heavy rains came at the wrong times for most commodity farmers. But Satzewich said that he did okay, thanks to the diversity of his crops and their locations. "I've got carrots in storage that I'm selling at farmers' markets right now." A varied crop list and a number of locations are his version of crop insurance. Some things get hit in some places, but other things make it and generate revenue. This is also his competitive advantage over other organic market gardeners who farm outside of the city and are more prone to frost and more reliant on rainfall for water.

Then our conversation took another interesting turn when I asked what crops he could grow that might surprise people, given Saskatoon's northern latitude. "Basically, we can grow everything here in the city that people can grow even far south." You just have to be smart about selecting early-ripening or fast-growing varieties. It's only when you get into really hot-climate crops, like sweet potatoes and watermelon, for example, that require a very long growing season that things get "tricky," said Satzewich.

"But not impossible?" I fired back.

"No."

But surely the drastically short growing season in Canada is a problem, I asked. Satzewich conceded that it was "an economic hindrance," but not as much as a home garden would assume.

There's an unwritten rule in Canadian gardening that you don't dare put any plants or even seeds in the ground until the third weekend in May because of the high chance of a late-spring frost. "Part of the whole SPIN farming mindset is breaking away from the home gardening practices! I'm planting as soon as the snow melts, early in April. I fall-plant crops like spinach," explained Satzewich. Spinach, lettuce, green onions, and peas for pea shoots are very frost-tolerant plants. Farming, as opposed to home food gardening, is all about knowing what you can get away with. "Farmers have to push the envelope to make a living." And when you're farming in the city, you can push the envelope a little beyond what you can in the countryside. "You have your microclimate as one of your advantages."

There are even advantages to being in a colder climate. The cold and dry winters on the Canadian prairies keep the pests away for the most part. One would-be farmer who attended a SPIN farming workshop was surprised that Satzewich could grow summer squash organically. In Oklahoma, the attendee explained, it's impossible to grow squash organically, given the overwhelming pests.

Easy access to water is another urban advantage. Satzewich doesn't have to wait for irrigation canals or for streams to thaw. That's another advantage that he has over fellow market gardeners who still grow outside of the city.

In 2001, Roxanne Christensen, a writer in Philadelphia, Pennsylvania, was interested in the economics of urban agriculture in her city, but she had no farming experience. As it turned out, the economic development director of the Philadelphia Water Department was also interested in urban agriculture to prove that economic activity could be generated while at the same time having a positive effect on the city's environment. (The successes of many environmental initiatives are difficult to document because, really, they are about the damage that doesn't take place. An urban farm's productivity can be measured economically, on the other hand.) Christensen had heard about Satzewich's sub-acre urban SPIN-farming model and felt that he could provide both the business model and the practical knowledge. Satzewich agreed to consult on the farm's plan, and the Somerton Tanks Farm broke ground on a half-acre (two thousand square meters) of city land in the spring of 2003.

Christensen had to establish a nonprofit organization to comply with city regulations, which prohibited leasing municipal land to private individuals and private companies. Her sales goal for the farm's first year was modest—$25,000 in gross revenue.

"The agricultural experts told us that was impossible," recounted Christensen by phone in the winter of 2010.[14] Universally, she found the whole "ag status quo" was skeptical. They trotted out the sales figures for conventional multi-acre farms—the pinnacle of agricultural efficiency—as a yardstick. The gross for a half-acre on a rural farm was $3,000 for a half-acre, at best.

But following the suggestions of Satzewich on which crops to plant, Somerton Tanks Farm grossed $26,000 the first year. Furthermore, the revenue climbed steadily over the next three years. By 2007, the farm generated $68,000. "And we hadn't even experimented with stretching the season," Christensen emphasized.

The timing was good. Farmers' markets were gaining in popularity, local food awareness was growing, and "people were willing to put their money where their mouths were," said Christensen. She's adamant that what ultimately makes urban farming a viable business is whether there is a clientele willing to pay for local food and markets at which to sell the produce. The point had been made, given a $68,000 seasonal income off the Somerton Tanks Farm location. SPIN farming, as a business model, had been proven.

The problem, as Christensen saw it, was that this system, though proven, was all "in Wally's head at the time." Farming traditionally has been a trial-and-error or, slightly better, a long, hands-on apprenticeship scenario with an experienced farmer. If Satzewich had to consult personally with every urban farm that wanted to use his sub-acre model, or if they had to trail around with him for hands-on learning, the uptake would be slow. But Christensen and Satzewich realized that this model was already somewhat systematized; it just needed to be stated into words. "It was very franchise-ready."

Christensen and Satzewich struck a business partnership agreement. He would be the farming expert who would continue to innovate and tweak. She would be the writer who could put the system into incremental instruction guides, accessible for novice urban farmers.

"We characterize it as a self-serve, self-directed, online learning series for self-starters," Christensen said. The basics of SPIN farming are detailed in seven modules, each guide available for purchase online for $11.99. The bundle of seven is $83.93 and can be purchased as a print textbook. When I remarked that it was a small sliver of the cost of an agricultural degree, even in Canada, Christensen quickly pointed out, "And those are largely outdated techniques and methods" being taught to young farmers.

Clearly, Christensen and Satzewich's hard work has paid off. There are now over six hundred practicing SPIN farmers who share successes and failures, and who support one another on an e-mail discussion service. Interest in the SPIN model continues to grow, but Christensen was quick to remark that the interest was strong early on, even in 2003. Most entrepreneurs coming to SPIN have no farming background or any practical experience. They are drawn by the economic potential. And Christensen is quick to point out that this interest predates the economic crisis of 2008. People were, and are, attracted to SPIN farming because of the opportunity, not because of desperation. "Farming is no longer viewed as a downwardly mobile profession."

One of the uphill battles of urban farming is that, in many cities, it is at the mercy of outdated municipal policies. In Philadelphia, it's still illegal to sell from the same land that you grow on. And there's really no smooth process that helps entrepreneurs to get access to land in the city, but Christensen gets calls "daily" from other municipalities that are looking for a boilerplate policy and paperwork to cope with the volume of requests coming in from would-be urban farmers.

Only the markets will decide whether urban agriculture will be a lasting business concept in places like North America. "If these ways really are better, they'll dominate. We don't have to fight anything, we just have to do it," Christensen replied when I asked her if SPIN farming would be around in a decade or two in North America. The "new frontier" aspect is in the form of interest from people in developing nations who can benefit from a relatively easily adoptable business model for very little cost—even in those countries whose governments are desperately trying to move their populations away from farming to get more in line with industry in the developed world. If people in the developed countries like Canada and the United States are choosing to be involved in urban agriculture, Christensen reasons, then it might become an acceptable option for entrepreneurs in developing countries. Christensen was particularly excited about the possibility of helping women in Sierra Leone establish SPIN-farming businesses. As a brand, SPIN farming provides an instant professional identity. "If we can export Pepsi®, we can export SPIN farming and have it accepted and have it become a 'force multiplier.'"

*   *   *

At the height of his SPIN-farming productivity, Satzewich had twenty-five urban plots, ranging from 500 square feet (46 square meters) to 3,000 square feet (275 square meters).[15] He's since cut back. SPIN farming and revenue from his SPIN-farming workshops and guidebooks that he and his now–business partner, Roxanne Christensen, have put together have put him in a position that many farmers can't even contemplate. He's thinking about his retirement plans now. Satzewich and Vandersteen have bought some land in a small town that Satzewich says is like a "Clint Eastwood movie set." But he's already growing some market crops out there, so retirement might just mean something different to a farmer than it does to the rest of us.

Chapter 10

TORONTO

Cabbagetown 2.0

The point of cities is multiplicity of choice.
—Jane Jacobs, *The Death and Life*
of Great American Cities, 1961

Toronto, Canada's most populous city, has two nicknames that speak to its food roots: "Hogtown" refers to Toronto's early prominence as a pork-packing hub in Canada (much like Chicago was in its early days), and "Cabbagetown" is another working-class moniker given to the eastside neighborhood where Irish immigrants grew cabbages in their front yards in the mid-nineteenth century.

Toronto is still bursting with food gardens, but these days those front yards, community gardens, and allotment gardens are just as likely to contain okra, Thai basil, and tomatillos as they are to sprout heads of cabbage. Toronto is Canada's most culturally diverse city, and the city's backyards and gardens are a United Nations of ingredients. Forty percent of Torontonians grow some sort of food at home, and countless others grow food in the 226 known community gardens spread out across the city.[1] (Toronto has a number of unofficial, self-organizing "squatter" community gardens, run by groups that have simply commandeered vacant and unoccupied areas for unofficial community gardens on land that is not part of this total.) The city also operates twelve allotment gardens totaling 1,674 individual plots, but there's a waiting list of five hundred people hoping to get an allotment plot.[2] If the capacity were there, Toronto would be virtually dripping in food gardens, some of them established by choice as leisure pursuits and others established by necessity.

FOODSHARE TORONTO

I was in Toronto, arguably Canada's food capital, on a postcard-perfect clear October day. The air was crisp and thin, but the sun was hot. I had made my way to the Ontario provincial legislature grounds known as Queen's Park for a harvest celebration called Eat-In Ontario. As I approached, it looked as if a mammoth food-focused circus had just rolled into town.

I watched in amazement as a group of fourth-grade boys huddled in a tent cheerfully sampling beets—albeit in dried, chip form. According to the chef who had prepared and was serving the beet chips, there had been beet-chocolate cupcakes swirled with ruby-red beet-infused cream cheese icing earlier. But the previous gang of ravenous elementary school kids had wiped out the supply before moving on to the "garlic tent," where another Toronto chef was piping garlic foam onto crackers for the kids. Under dozens of tents, groups of children of various ages were paying close attention as they learned about local vegetables and produce, and actually enjoying samples of foods like squash and carrots.

After the roving herds of children had grazed for about an hour, organizers gathered them up for the pinnacle event of the day. The kids sat on the grass and squirmed in sheer anticipation as people in bumblebee outfits and dressed up as sunflowers led the countdown. Then, at 1:45 pm exactly on the command of the bee and sunflower people, several hundred kids brought giant red apples to their lips and took the biggest bite they could manage. It was all part of the Big Crunch 2010—a synchronized bite shared by 64,000 kids from 125-odd schools across

Canada that day. The Big Crunch 2011 doubled its participation with 112,352 crunchers who simultaneously shared a bite from coast to coast.

FoodShare, the originator of the Big Crunch and the organizer of today's harvest event, was founded in 1985 and is now Canada's largest community food-security organization. From the beginning, it has been a proactive food-security organization that uses community-development strategies as alternatives to the usual stopgap emergency food-relief services provided by food banks. FoodShare began a subsidized CSA program called the Good Food Box in 1994. Community gardens, urban-agriculture demonstrations in low-income neighborhoods, and food instruction via community kitchens were part of its earliest programs, filling the "food-literacy" gap by passing on knowledge that used to be taught in the home and on the farm.

FoodShare has grown into a web of educational and community outreach programs and an advocacy and activism hub for food justice in Toronto. With its 682-square-foot (63-square-meter) demonstration urban garden and compost facility, it has helped to facilitate 720 student nutrition programs as part of its role as a key partner in the Toronto Partners for Student Nutrition, and it provides dozens of different school food-literacy programs as part of its Field to Table Schools program. It offers mother-and-toddler nutrition classes. FoodShare also operates seventeen "Good Food Markets"—farmers' markets in the low-income neighborhoods that are not only poor candidates for farmers' markets; they arc urban food deserts.

"Toronto is the most developed and complex—in a food sense—city in the world," Debbie Field, executive director of FoodShare, declared to me as we chatted among the hubbub of the Eat-In Ontario festival.[3] This makes FoodShare's seemingly simple mission—to work with "communities to ensure that everyone has sustainably produced, good, healthy food"—all the more tricky.

Toronto is Canada's most culturally diverse city, and it's one of the world's most multicultural cities per capita.[4] It's a huge immigration hub, absorbing 55,000 new arrivals each year.[5] Fifty-one percent of Toronto residents were born elsewhere. (The Peoria, Illinois-born, New York-raised Field herself is part of this demographic.) Just considering the multicultural makeup of the city, not to mention its population of over 2.4 million residents, has a huge impact on how you think about food security, explained Field. So, while FoodShare works to build a resilient food system, Field knows that "in a city of immigrants, it's *as* important to have mangoes [available] as apples." And because of the immigrant population—a sector of the population that often values fresh produce much more than Canadian-born eaters—you have a high cultural demand for fresh and varied produce.

Toronto, like any big city, has food deserts. Field cited a recent study by Darcy Higgins, a Toronto-based food activist and writer, who researched the proximity of grocery stores to residential areas in Toronto and found that only 51 percent of residents in the city live within 1 kilometer (0.6 mile) of a grocery store.[6] "So 49 percent of people are five, ten, fifteen kilometers away from anything," Fields reemphasized.

The lack of planning on the part of the city to include food retail in zoning and planning discussions is currently one of Field's hot-button issues. Field explained that "FoodShare wants to change the city's planning act so that developers can't build housing unless they can show that someone can access fresh food from within one kilometer of where they live. You can force them to do it in a mini-moment if there is political will!"

At first this seemed a bit extreme to me, until I recalled that I grew up in a community that had that family-owned-and-operated grocery store with the produce and fresh meat section, and it was only four blocks from the house.

"In some neighborhoods, people can't even afford the twelve-dollar [Good Food Box] cost," said Field, of the FoodShare's home delivery of weekly fresh, local produce. So recently, FoodShare has begun to establish Good Food Markets in low-income neighborhoods. Good Food Markets are simply a table of fresh products operated by two women in the lobby of their low-income neighborhood housing complex or in a nearby parking lot. The sales don't amount to much—about $300 worth of food on "market days," but the project accomplishes the task of getting fresh, culturally appropriate food into neighborhoods that don't have the income potential needed to attract farmers' markets. "A farmer won't go into a farmers' market unless he or she can sell $1,000 per day," Field explained, which is why low-income neighborhoods rarely are good candidates for farmers' markets.

FoodShare is also heavily invested in the potential and payoffs of urban agriculture. It not only operates its demonstration urban farm and composting center at its headquarters in central Toronto, but it also supports urban-agriculture programs at addiction-treatment facilities, mental health centers, and schools. "I mean, right here, shouldn't there be corn growing right now?" Field asked seriously, as we contemplated the enormous manicured lawn surrounding the provincial legislature building.

Field then acknowledged that while the complexity of the city's food system poses challenges, it is also complex and developed in a good way. Notably because of the forty-acre (sixteen-hectare) Ontario Food Terminal, located on a major roadway in the city. "Toronto is one of the only cities in the world that has a provincially funded location where local farmers can go, and where everyone from small vendors to chains can come and get food." Toronto is adjacent to the world's largest protected agricultural zone, the 1.8 million acres (728,400 hectares) of prime farmland and woodland called the Ontario Greenbelt.[7] The Ontario Food Terminal is a key element in facilitating the movement of local food grown in the Greenbelt to customers while earning farmers a living wage.

Indeed, the Ontario Food Terminal is one of the few noncorporate-owned major urban food-distribution hubs. They have all but disappeared because multinational corporate grocery chains have their own "supply management hubs" that effectively shut independent grocery retailers and smaller farming operations out of the market in cities without an open wholesale hub. Operated by the Ontario Ministry of

Agriculture as a small city-within-a-city, the Ontario Food Terminal is where area farmers can bring their fruit and produce to market and sell directly to retailers and restaurants. It competes directly with other fresh produce brought in from around the globe, but the fact that there is a terminal in which produce can be sold in a central location to buyers has allowed many smaller farmers to continue farming and has also enabled consumers to continue accessing local products through traditional grocery markets. This rarely happens anymore.

Field noted that in her Toronto neighborhood, her favored independent grocers and chefs all get up at four or four-thirty in the morning almost daily to get fresh local produce that has come directly from nearby farms. While Field's central downtown neighborhood has around twenty greengrocers, not every area in Toronto is as lucky.

INDIVIDUAL, COMMUNITY, AND SOCIETY

Toronto has thirteen "priority neighborhoods," as they are called—identified for their higher levels of social risk factors such as low-income, high-unemployment, single-parent families and recent immigrants.

As Field sees it, lack of income to purchase food is only a third of the problem in "our unsustainable, dysfunctional industrial food system." There's a global "food and income crisis" where "1.2 billion people will go to sleep tonight hungry and undernourished from lack of money." Most people, Field explained, intuitively understand this part of the food system. No money equals no food. And this equation is what the traditional food-bank model focuses on.

Field has also identified an equally destructive but less acknowledged problem: the global "food and health crisis" in which "for the first time in human history 1.2 billion people will go to sleep malnourished from overnourishment of the wrong kind of food! Childhood obesity and diabetes will overcome every government's budget in the world." She then remarked on the other side of this tragedy, the ongoing farmer suicides that occur because farmers can't make enough money to live on while growing these commodity crops that feed into our industrial food system—the very same system that is killing people with "diseases of affluence."

"In Canada, right now, [Prime Minister] Stephen Harper has absolutely no idea how we would eat if the border between the United States and Canada closed, or the border between the United States and Mexico." One of Field's political ambitions is to lobby the Canadian government to create a Ministry of Food Security. "In New York, during 9/11, there were *four* days of food in the city. Total." Field explained that she hates the term *food security* because "it's war-based," but she conceded that it's not a bad way of getting at part of the vulnerabilities of a food system.

"We need a Minister of Food Security. Neither Stephen Harper nor Barack Obama get up in the morning and think about how the people in their country are going to eat the way a woman thinks about what she's got in the fridge and what she's going to serve," said Field. Part of our current food crisis challenge is the lack of representation of women "who remain connected to traditional attitudes toward food," explained Field. She is not advocating for women to hunker back in their homes and stay out of sight; rather, Field contends that we need women in the political and social spheres who truly understand how food systems work on an individual, community, and social level.

The final element of Field's social food policy, the one that would address the "food and agriculture" crisis, is controversial, she admits. Field would like to see subsidies for farmers who produce for the local market. Even farmers balk at this idea, Field says, because their opinions are based on the US subsidy system. But she cites successful farming subsidies on certain items in Kerala, India, which have been in place for thirty years. "We could have five things in Ontario, like cheese, carrots, apples, soy nuts, and broccoli, that could all be twenty cents per kilogram in all the grocery stores. But the farmers could get the full [wholesale price] amount. So this is the policy we're working on right now."

As for the Big Crunch and Eat-In Ontario, it's no coincidence that it is being staged on the doorstep of the provincial legislature building. FoodShare is lobbying for provincial and national child and student nutritional programs. FoodShare has launched its Recipe for Change initiative to bring food literacy to schools so that all students can learn how to make healthy food choices, access at least one healthy meal a day at school, and increase their physical fitness through food activities such as gardening, cooking, and composting at school. Canada is one of the few Western nations that does not have a national school nutritional program or national school nutritional policy. Lunches and snacks have always been the responsibility of the student and family. And so, rather ambitiously, the work at FoodShare, said Field, is about "creating capacity" at all three levels and addressing all three crises at the human, individual level, the community level, and "eventually the social level."

THE STOP COMMUNITY FOOD CENTRE

FoodShare is not the only pioneering community-supported food-security organization in Toronto. In the mid-1970s, The Stop Community Food Centre opened in Toronto's West End as one of Canada's first food banks. Since then, it has evolved into a global pioneer of innovative antipoverty and food-justice programs, drawing visits from the likes of UK chef turned food-education crusader Jamie Oliver.

Like FoodShare, The Stop takes a holistic community-development approach from the point of view that food should be a basic human right and that empowering people to grow food and help one another helps communities build resilience. What would otherwise be termed a soup kitchen in a food bank facility is a bright, welcoming dining room at The Stop, where fresh, local, pesticide-free produce and organic beef is served to clients at the table rather than at the usual cafeteria-style lineup. The drop-in dining area transforms into a classroom for nutrition education and perinatal programs, as well as other community outreach programs.[8]

In 2009, The Stop expanded to a second location, called the Green Barn, which became the operational headquarters for its urban-agriculture programs in low-income and other socially disadvantaged communities. It operates three community garden sites in inner-city Toronto that together yield up to 4,000 pounds (1,800 kilograms) of fresh, organic produce a year. Educational workshops encourage peer-to-peer exchanges on how to grow fresh, local, culturally appropriate organic produce. And The Stop facilitates a Yes-In-My-Backyard (YIMBY) garden-sharing program that connects people who want to grow food with those who have the space.

In the summer of 2010, The Stop established its Global Roots demonstration garden at its Green Barn location. The Global Roots

The Stop Community Food Centre's greenhouse for year-round organic production and gardening education, Toronto, Ontario. Photo © The Stop Community Food Centre. Used with permission.

Garden features seven twenty-foot-by-thirteen-foot (six-by-four-meter) plots planted with foods from some of Toronto's most populous ethnic communities: Chinese, South Asian, Somalian, Italian, Latin American, Polish, and Filipino. Elders with often decades of food-growing experience from the various ethnic backgrounds do the teaching and directing, while the youth gardeners do the heavy lifting and digging. Together, these intergenerational teams tend these garden plots, socialize, and even cook together. The food grown is not just a demonstration of what can be grown in Toronto's climate and the diversity of food cultures that make up the city; produce from the gardens go into The Stop's community kitchens and wood-fired community baking ovens, and back into feeding the community.

Wayne Roberts

While in Toronto, I also met up with Wayne Roberts, another deep-in-the-trenches food-security activist and author.[9] Roberts had been agitating in social-justice circles around environmentalism and labor issues for a couple of decades before becoming the manager of the Toronto Food Policy Council from 2000 to 2010. Recently retired, he seemed busier than ever, writing widely on food-policy issues and serving on several boards, such as FoodShare, Community Food Security Coalition, and Food Secure Canada. His consulting and speaking work was increasingly taking him beyond Toronto and even Canada to consult on community food-security issues. Life after work seemed to be a busy next step for Roberts.

Roberts wanted to show me what he thought the future of Toronto might look like, and how that future was already taking shape in economically disadvantaged neighborhoods. Just a few blocks off the major routes of Bathurst and Dundas Streets, we were in a multiethnic neighborhood with one of the lowest average household incomes in the city but that was also dotted with food gardens. We stopped briefly at the

Alexandra Park Diversity Garden, tucked away along the flank of an open stretch of a municipal park. It was just minutes from Scadding Court Community Centre, which, on the surface, looked like a typical community center building built in the late 1970s. A community garden overflowing with bush beans, broccoli, Swiss chard, sunflowers, kale, squash, marigolds, and the like was where the lawn next to the main entrance would otherwise have been.

Roberts was overflowing with enthusiasm for the various urban-agriculture initiatives facilitated through Scadding Court Community Centre. He walked me through the various community gardens, the community compost education center, a community café that serves low-cost, healthy meals and provides job training for youths, and gushed about the food-literacy programs for kids. For a week every June, the community pool in Scadding Court Community Centre is converted to

Community gardens at the entrance of Scadding Court Community Centre, Toronto, Ontario. Photo by author, October 7, 2010.

a trout pond, stocked with live fish for the Gone Fishin' program. Children learn to fish and even take home their catch of rainbow trout, for a minimal fee.

Scadding Court Community Centre most recently established a street market along Dundas Street. A line of modified shipping containers have been refurbished into street market stalls, similar to what many of the residents of the neighborhood are familiar with from their countries of origin. This market, which includes food stalls, provides a lively street scene, especially in the evenings, and a small-business incubator for residents of the community.

Within a very short time and on foot, Roberts had taken me to see several community gardens as well as the Scadding Court Community Centre, with all its various community-based agriculture programs. This is what Roberts wanted me to see, what he hoped the future could hold for more neighborhoods in Toronto, especially those struggling with crime and social problems.

"Busy places, where there are lots of eyes," said Roberts, "is a safe place for a community." Urban agriculture isn't really about food, in Roberts's opinion; it's about building vibrant, safe communities with positive options, as opposed to the negative ones usually associated with inner-city life in North America. Roberts sees Scadding Court as a particularly positive place for inner-city youth in Toronto.

"I always say, 'Don't ask me if at fifty dollars a square foot, can you justify an urban-agriculture program in the city of Toronto. Ask me if at $100,000 a year per juvenile at a detention center, can we not pay for one urban-agriculture program instead?'"

TORONTO'S REBELLIOUS STREAK: GUERILLA COMMUNITY GARDENS AND OUTLAW CHICKENS

My friend John H. offered to show me around a community garden in Toronto that he had belonged to, as long as I was discreet about how I photographed it and wrote about it, so as not to give away its location.[10] The funny thing was, it was plainly visible behind a see-through metal mesh fence on a residential street surrounded by both low-rise and high-rise apartments. The only thing different about this community garden was that there was no usual sign with the garden's name, no poster of garden rules, and no contact information. And as far as the city was concerned, this garden didn't exist.

The space on a former industrial site leased from the city had been

Striped snail on an eggplant leaf, secret community garden, somewhere in Toronto, Ontario. Photo by author, October 7, 2010.

proposed as a community garden for men who lived in a nearby shelter, but the garden never really took off. With the infrastructure already there, gardeners from the neighborhood began to use the space anyway to grow food. They self-organized and now maintain the dozens of raised beds and plots on a piece of inner-city land that they technically don't have permission to garden on. As long as the city continued to be unaware of the garden or turned a blind eye to it, no one but the gardeners would know that it was not technically supposed to exist. Besides, it looked and felt like any other community garden to a passerby or casual observer.

The humidity off Lake Ontario was thick, even in the late fall. There were striped snails hanging from plants, prickly-pear cactus to watch out for, and a confusion of plants, pea fencing, rain barrels, and haphazard footpaths between overgrown plots. We had to scramble over places where weeds had overtaken a pathway or where rusty wheelbarrows were abandoned. On closer inspection, though, the gardens were still producing giant fountains of Swiss chard, broccoli, long green beans, herbs, eggplant, and exotics that were strange and unidentifiable to both John and me. I relished my moment in this garden that was hiding in plain sight of newly built condo towers encroaching on the decaying old red-brick factory buildings and warehouses of the area's previous lifetime.

*   *   *

One area of urban agriculture where Canada is strangely lagging behind the curve is in its bylaws for keeping urban chickens. The anonymous blogger behind the website Backyard Chickens in Toronto (http://www.torontochickens.com), known as "TC," lists eighty-six municipalities in the United States that allow city residents to keep chickens for eggs (with varying restrictions), compared to Canada's six.[11]

But scratch around in just about any city in the United States or Canada, and, regardless of whether it's legal to keep chickens, it's pretty easy to find backyards with small coops and a small brood pecking around. In 2009, *Raising Chickens for Dummies* was added to the wildly popular "For Dummies" book series, and urban chicken-keeping supply shops have been popping up in Portland, Oregon, and London in the United Kingdom. Hardware stores are starting to carry chicken supplies, and there are even underground mobile urban slaughterhouses that operate—very discreetly—in a handful of cities for clients who wish to put their aging hens in the stewpot when a hen's prime egg-laying days are done.

Toronto-based urban-agriculture advocate Lorraine Johnson published a book on urban agriculture titled *City Farmer: Adventures in Urban Food Growing* in 2010, in which she writes extensively about her three urban (outlaw) chickens.[12] I figured, correctly as it turned out, that she would be more than happy to show me her small brood of three laying hens. After I promised not to disclose her actual street address, Johnson happily invited me over to meet Nog, Hermione, and Roo.[13]

I arrived just in time for feeding. Johnson led me through her house and into a long, narrow backyard that was being overrun with pumpkins the size of beach balls. Passing the garden shed and stepping warily over the thick pumpkin tendrils snaking across the grass, we arrived at the wire fencing and gate that enclosed an inner wire cage for extra predator proofing. This double-layered security, Johnson assured me, was "what you want in the city." Raccoons are the main predator concern for urban chicken farmers.

The hens scurried and chortled as we approached but quickly resumed their pacing and pecking. Nog, the dark, small Babcock chicken almost black with iridescent auburn streaks, and Roo, the ginger-feathered Buff Orpington, looked like they might attempt a run at backyard freedom, but Hermione, a snowy-white Araucana with red trim, just strutted back and forth in front of her modern sky-blue "Eglu" prefab hen house, manufactured by UK backyard urban-agriculture product company Omelet.

Johnson tossed a few handfuls of feed pellets next to some vegetable peelings, and the day's chores were pretty much done. Every second day, Johnson cleans out the Eglu's waste tray under the coop and changes the bedding and straw. For relatively little effort, Johnson said that on average, she could count on eighteen eggs a week between the three hens.

I remarked that I'd been in several commercial egg-laying operations with thousands of birds in cages whose shrill calls would compete with the loud buzzing fluorescent lamps and giant fans that move the ammonia-sharp air around. *That* had been my impression of keeping chickens up to that point, but Johnson's three-hen setup was nothing like that. There was no smell, other than that of the straw bedding and the mulching leaves from the overhead trees that provide shade and shelter. So much for controversy. Even Johnson laughed at the nonevent of it all.

The most common reason people keep chickens in the city is for the

Hermione, one of the city's outlaw urban chickens, strutting in front of her designer "Eglu" coop, in a backyard somewhere in Toronto, Ontario. Photo by author, October 8, 2010.

ultra-fresh, tastier-than-store-bought eggs. Also, chickens are keen consumers of just about any type of kitchen peelings and act as roving composters and fertilizers for urban farmers. But concerns about animal welfare, predators like coyotes and foxes being attracted farther into cities, noise pollution, and smell are common arguments against allowing urban chickens to become backyard fixtures. For a while, avian flu concerns were an issue as well, but they seem to have faded in the wake of the more recent swine flu panics.

Toronto is currently grappling with the pros and cons of changing its bylaws to allow small broods of chickens in residential backyards, just as many cities are faced with a de facto urban chicken movement on their hands. Johnson's three hens figure prominently in her book and in her magazine and newspaper articles. Toronto, for instance, already has a well-publicized Toronto Chicken Coop Tour, much like the regular coop tours that take place in Portland, Oregon, or Madison, Wisconsin. The people who go on the Toronto Chicken Coop Tour are asked to keep the locations of the outlaw coops to themselves. And the list of urban chicken blogs continues to grow by the day in Toronto, as it does in most cities, with or without the blessing of city hall.

NOT FAR FROM THE TREE

Laura Reinsborough, founder of Not Far from the Tree, admitted that she hadn't picked a piece of fruit from a tree until she was twenty-five and volunteering at the Spadina Museum in downtown Toronto.[14] But oh, what an epiphany it was. At the time, she was a graduate student in environmental studies at Toronto's York University and had "experience in urban agriculture," so she was asked if she could pick the ripe fruit from the museum's fifty-some apple trees on its grounds. The Spadina Museum had the capacity to maintain the heritage orchard but couldn't deal with the actual fruit production that resulted. Reinsborough agreed to pick the apples—the first of many revelations of the caches of fruit hiding in plain sight in an urban setting.

"The whole city just changed for me," Reinsborough said, her voice still rising enthusiastically at the thought of it. "I became so much more aware of what was growing in my neighborhood and the potential of urban soil to produce amazing, delicious . . ." Her voice trailed off and then rebounded. "It's not just that there's fruit, but there's an abundance!"

In 2008, Reinsborough put her master's degree into action when she mobilized a dozen of her friends to collect the sour cherries, apricots, mulberries, apples, and other fruits from her neighborhood that were otherwise just falling to the ground and rotting. Reinsborough had approached the trees' owners and offered to bring a crew to pick the ripe fruit for free. The deal was that they'd give one-third to the trees' owners, keep one-third to divide among the volunteers, and donate one-third to local shelters or community centers for distribution to residents who didn't have access to fresh fruit. The bonus was that these backyard trees were mostly unsprayed—the crew avoided fruit grown on any contaminated ground—and therefore mostly organic by default.

Reinsborough describes her way of looking at her city now as "putting on her fruit goggles." She started to see how the potential for urban fruit production lay right under her nose once she actually started to pay attention. She noticed fruit trees in ravines, in city parks, and on private property. Stringing them together was not so much like an urban forest but a giant urban orchard. "The whole city changed for me," she remembered.

She realized that her immediate neighborhood was a bonanza of free, fresh, organic urban food and promptly launched Not Far from the Tree as an urban fruit-gleaning nonprofit with an entirely volunteer workforce, of which she is a part. By the end of 2008, Reinsborough had a corps of 150 volunteers who picked 3,003 pounds of fruit off just 40 residential trees. In 2009, Not Far from the Tree had 450 volunteers, with four paid staff, thanks to grant money and private donations; and had collected 8,135 pounds from 125 trees, donating a full 2,711 pounds to local charity groups.[15]

Urban fruit gleaning is not unique to Toronto. The number of fruit-rescue outfits has exploded in cities everywhere, enabled in part by social media and the Internet, where it's easier to connect people with unwanted or overproducing private fruit trees in cities, with the teams of volunteers ready and willing to pick and redistribute the bounty. Most major cities now have fruit-gleaning groups at work. FoodForward in Los Angeles has picked 434,269 pounds (197,000 kilograms) of fruit in urban Los Angeles since its inception in 2009 and has given 100 percent of it away.[16] There's also the Portland Fruit Tree Project, which began in 2006 and in 2010 collected 29,397 pounds (13,334 kilograms) of fruit that would have otherwise gone to waste.[17] Food banks across North America have even begun to organize their own gleaning crews.

In 2010, more than 700 volunteers with Not Far from the Tree picked 19,695 pounds of fruit from trees in just a handful of residential neighborhoods in Toronto. That said, they got around to only one-quarter of the fruit trees that Toronto residents offered up via the group's online tree registration on its website. In 2011, Not Far from the Tree launched a public fruit tree–mapping initiative to get a better idea of just how much fruit grows in Toronto. The fruit picked in 2010 is just a fraction of the 1.5 million pounds of tree fruit that Reinsborough estimates ripens in the city every year.

Not Far from the Tree recently started a tree-tapping pilot project, called Syrup in the City, to tap trees like Norwegian maples and other urban sugar-producing trees. Other partnerships with community-based food groups have been made to offer cooking and fruit-preservation classes to fuel the enthusiasm that seems to be building for urban fruit production and gleaning.

Reinsborough is also consulting with other communities and cities about starting fruit-gleaning programs, and the requests are coming in from all over North America. She has even received queries from Scotland

and Puerto Rico. But her ambitions are really focused on the task close to home. She hoped to expand into fourteen or fifteen different neighborhoods in 2011, but she quickly noted that there are a total of forty-four neighborhoods of that scale in Toronto. Reinsborough originally cobbled together funding from provincial and municipal government sources but now is shifting that toward private funding and other revenue streams. Undaunted, she is convinced that soon Not Far from the Tree will cover the city. "We see ourselves getting there within a few years."

Growing Public Orchards

It's infuriating when deer maraud through the pea or corn patch and nibble it down to stubs. But nature can be forgiven; the human element, not so much. Produce-napping and unauthorized gleaning in any community garden or urban farm is no more welcome than vandalism.

Talking to community gardeners and garden managers, stories of thieves who made off with the ripe tomatoes, pulled beets, pinched herbs, and stripped raspberry canes were easy enough to come by. Most urban gardeners at community gardens just accept this as part of urban living, and they plant a bit extra to compensate for it. (Anecdotally, urban gardeners have told me over and over again, the culprits are usually "little old women"; the second-most common tomato-nappers or herb pinchers are fellow gardeners.) Often, if the problem persists, gardeners collect fees from the collective for fences and padlocks.

But what if foraging and grazing wasn't just expected but encouraged? What if parks were places to go for a walk on a sunny afternoon where you could pick an apple off a tree and crunch away in plain view of others? This is exactly the idea behind municipal public orchards and other edible landscapes that some cities are experimenting with—for example, a nine-fruit-tree public orchard in Ben Nobleman Park in Toronto, and others farther west in Calgary, Alberta.

On a mid-July day, I sought out a public orchard in Calgary, a city better known for its oil- and gas-industry corporate offices than for forward-thinking municipal food-security initiatives. But in this case, Calgary was ahead of the curve in Canada, having established three municipally planted orchards in 2009 as part of a five-year test pilot. The idea was to test management models—city-run orchards to self-organizing community-run orchards—to gauge which type would be more effective. The city also needed to gather information on which fruit and nut trees would grow; how much they would produce; what pest and disease pressures they could withstand; how well they worked in the larger community parkland; and, at the most basic level, whether orchards could even get a foothold in a city where there are often fewer than a hundred frost-free days per year.

I decided to visit the Hillhurst-Sunnyside Community Garden orchard, the first completed community orchard in the city. The com-

Hillhurst-Sunnyside Community Garden and adjacent community orchard pilot project, Calgary, Alberta. Photo by author, August 14, 2010.

munity is one of the city's oldest residential areas, but it is also one that has resisted densification despite being separated from the huddle of skyscrapers at the city's business core only by the narrow Bow River. The orchard was at the end of an inner-city residential street of neatly kept single-family homes. The first thing that struck me was that it was not an orchard as I had pictured in my mind (well-organized straight rows of mature fruit trees). Instead, it was a haphazard collection of saplings

Apple tree at Hillhurst-Sunnyside Community Orchard Pilot Project, Calgary, Alberta. Photo by author, August 14, 2010.

planted rather close together on a triangular point where a small park with spruce and birch trees rimmed an irregular patch of lawn.

A steep hillside rose behind the orchard, intensifying the summer heat. This, I thought, would hold the orchard in good stead. (Clearly, the settlers who named Hillhurst and Sunnyside were literalists.) But as I walked closer, I could see that the effects of a recent mid-June hailstorm had pockmarked the small green apples and pears. Other saplings had been almost eaten down completely by deer. The apple, pear, hazelnuts, apricot, and cherry trees would have to fight hard to survive.

But it was possible to grow all these fruit and nut trees in cities, even as far north as Calgary and Edmonton, with latitudes of fifty-one degrees north and fifty-three degrees north, respectively. In fact, in the small park adjacent to the triangular Hillhurst-Sunnyside orchard, a couple of older apple trees had fruit hanging on them. A trio of teenage boys were kicking around a soccer ball and bantering among themselves. As I poked through the younger fruit trees, the boys stopped their game to each pick an apple from the older tree and began to talk about how delicious the (unripe!) apples were. After about five minutes, they whipped their apple cores into the woodchip beds that surrounded the grass and resumed their informal soccer match.

There was the proof that would convince any naysayer as to the value of a community orchard. If you plant it, they will come. Even teenage boys would enjoy fresh fruit.

I recounted this story to Tim Kitchen, president of the Hillhurst-Sunnyside Community Association, a volunteer position. He's also the orchard's steward in the community, which he admitted didn't amount to a lot of work. "The wonderful thing about trees is that you don't really have to do much. There's some really minor pruning, but the rest of the time, you just have to harvest."[18]

The city, according to Kitchen, organized a pruning workshop for the community, but at this early stage, the upkeep has been minimal. For now, the community's role is simply to monitor the health of the trees and liaise

Hillhurst-Sunnyside Community Orchard Pilot Project under snow, Calgary, Alberta. Photo by author, December 9, 2010.

with the city if a problem arises. "And the fruit disappears whether we organize a harvest or not," laughed Kitchen. "It would be nice if there was enough volume that we'd have to organize a harvest, but we're not there in terms of the trees' maturity yet." However, this self-organizing harvesting is the "best model going," as far as Kitchen is concerned.

The pilot project has been an education for both the community and the city. The trees were planted too close together, Kitchen said, and the location was not all that appropriate in hindsight. Kitchen acknowledged that this has been a quantum leap for the city government and its parks department. The city has had to respond to the public pressure by certain groups in the city to grow food on city land. "At the heart of it, agriculture is traditionally the responsibility of the province or state, not the municipality." The city, Kitchen suggested, has had to weigh this pressure with the risks of liability of growing food on public land for public consumption.

The new model that Kitchen felt was emerging as the most

promising was one in which the city established fruit-bearing trees associated with established community gardens. I would later learn about and witness this as the most common model at work in the United Kingdom as well.

PUBLIC EDIBLE LANDSCAPES IN THE UNITED STATES

Darrin Nordahl first noticed public edible landscapes while living in Berkeley, California.[19] Small veggie patches could be found on what was clearly public land, such as the areas between the streets and sidewalks. Sometimes they were planted with fig, orange, and Meyer lemon trees, something even Berkeley wouldn't have conceived of as part of municipal landscaping. Nordahl even noticed little vegetable gardens on the narrow strip between the curb and the sidewalk, a space so minimal and close to the road that he would never have considered it as a food-production space. On the odd opportunity he had to speak with these guerilla growers, he'd mention that they had planted on public land and that the produce was technically public produce. The growers never seemed to mount much of a counterargument. Usually, they'd simply agree with him.

Nordahl now creates exactly those types of edible landscapes as a city designer at the Davenport Design Center, a division of the Community and Economic Development Department of the city of Davenport, Iowa. He's become a Pied Piper of the public orchard and municipal food-policy movement in the United States since his book *Public Produce: The New Urban Agriculture* was published in 2009.[20]

The reality of fruit trees, much like zucchini, Nordahl explained to me over the telephone, is not the scarcity of produce but rather the abundance. "The reality is that people just want to be able to go out there from time to time and grab some fresh fruit for themselves and their family. They have no desire to eat the two hundred pounds of fruit that the tree is going to produce."

In *Public Produce*, Nordahl gives the example of Des Moines, Iowa, as proof that municipalities are starting to consider the benefits of growing edible landscapes on city land. The Des Moines municipal government's parks and recreation staff has planted fruit and nut trees around schools and community buildings. The maintenance on fruit and nut trees, some might argue, is even less than that of a lawn-carpeted park with ornamental shrubs and hedges. But the edible landscape provides a little bit of food security and teaches people about what types of foods are actually feasible in a certain climate. The department has even planted papaw (also known as paw paw; a tropical fruit similar to papaya), spicebush (an aromatic bush mostly native to eastern Asia), and serviceberry (a small purple fruit also called saskatoon berry in Canada), which ironically are novelties to most area residents but are actually native Iowan plant species. Part of the idea is to increase residents' food literacy as well.

Under Nordahl's purview, Davenport has also allocated $370,000 for reconstruction of an underutilized surface parking lot, which will soon become a "green space filled with fruit and nut orchards, garden plots, and pergolas replete with rambling grapevines."[21] Volunteer maintenance support will come from the local schools and university and from community groups such as Big Brothers Big Sisters and the United Way. And reportedly, the Thai restaurateur across the street is looking forward to planting, tending, and harvesting some Thai chili peppers, basil, and eggplant for his restaurant.

Nordahl lamented that tomatoes, that poster child for local produce, are difficult to grow well in Davenport. "It's best to leave that to the professional farmers," he conceded.

It doesn't take one-acre patches of land to produce food, Nordahl clarified. "Municipalities are the largest landowners within any city." Nordahl listed the inventory of municipal land that most cities have in abundance, all of which can be turned into more productive space: street medians; flood easements; utility corridors; parks; town squares; and the grounds around libraries, courtyards, schools, and city hall buildings. These areas typically are already planted with ornamental plants, so switching over to edible plants wouldn't really mean more up-keep. Just more output.

"The idea is not to rip out fountains and park benches but to start searching for spaces that you know will never be utilized," said Nordahl, knowing that citizens value their parkland, soccer fields, and other green space we generally think of as parkland. Cities own a lot of land that will simply never be developed because it's along a street, it's a utility corridor, or it's on a flood plain along a river.

Nordahl also took issue with the current solution for urban blight, which was to turn vacant lots into community gardens and other food-growing spaces. Community gardens on vacant land are especially susceptible to the economics of the situation. Once a vacant lot gets cleaned up and turned into a fertile space where food starts to grow, it becomes desirable again, and someone will want to develop it. Cities are always looking to increase their tax base, "and most of the time the community will lose." For true food security, these edible landscapes need to happen where the risk of competition from development is lowest.

The other problem with community gardens, Nordahl argued, is that they really aren't all that public. It's one thing for a private citizen to plant on public land in a community garden. "Community gardens, however, are tended by private citizens and are treated as small pieces of private land." The produce is not publicly available. "That's the crux of my book and where I am now personally as a government official. It has to be on public land, and it has to be available to the public," insisted Nordahl. And this is where Nordahl said the rubber hits the road in mainstreaming the concept of public produce: "These initiatives have to be funded, spearheaded, and supported by public officials and public policy."

"Some of the critics of my idea argue that this is not the role of government. I'm arguing that it is." Nordahl has seemingly hit the trend of public food growing on an upswing. Since his book has come out, he's become aware of the surge of other municipal governments taking a closer

look at edible landscapes instead of ornamental shrubs and plants. That is
what made former San Francisco mayor Gavin Newsom's 2009 food-
policy efforts to bring edible landscapes under the city's umbrella such a
bold move. Nordahl pointed to this as the way that local municipal governments need to be heading. "His directive was so clear that food-system
planning is the responsibility of local government," Nordahl said.

Most city governments that are leaning in this direction, however,
are only just beginning to do a land inventory. Often it's not even known
how much municipal land is available or appropriate for producing
food. Some cities are also including rooftop areas in this inventory.
Some estimates state that one-sixth of the landmass of most cities is covered with buildings with flat roofs—schools, big-box stores, high-rises,
malls, and warehouses—largely unused space that could be put into use
for food production.[22]

Nordahl acknowledged that the demand for fresh produce is very
high, hence the fenced and padlocked community garden plots. There
are even some community gardens where the individual plots have their
own fences and locks to protect from intra-garden raiding.

In fact, the question that Nordahl is most often asked is how to control the distribution of this public produce. Herein lies the beauty of
public produce, Nordahl replied. The food is available to the public, so
the situation has to be self-regulating, often at the community level. But
it usually, miraculously, works out in a rather civilized manner. The solution to the distribution problems is really as simple as not fencing off the
food and trying to control it. Rather, it's better to just plant more of it.
Make more of it available, and lower the scarcity issue.

Nordahl, though biased, sees this edible streetscape revolution continuing to spread, city by city, from a municipal level reaching up to a national
level. But the momentum will come out of the cities that choose to address
their food securities and not wait around for national food policies. As he
told Diane Rehm while being interviewed about his book on National
Public Radio: "City Hall will always act faster than Capitol Hill."[23]

Chapter 11

MILWAUKEE

Growing a Social Revolution

To truly change our food system, we need 50 million new people growing food in their local community.

—Will Allen, urban farmer and CEO
of Growing Power, Inc., 2011

Cities like Toronto, Vancouver, and New York get a lot of press applauding their urban-agriculture initiatives. Yet, it was in the American Midwest where I met actual hands-on innovators and saw the most ambitious projects at the forefront of the urban-agriculture movement. Chicago is on track to being the home of the world's first vertical farm, growing food on a smaller footprint by layering indoor fish farming on one floor, pigs on another, and vegetables on another, and so on—a farm in a skyscraper, in a sense. The world's largest commercial urban farm is being planned in Detroit, Michigan. And though he'd probably be uncomfortable with the title, a sweatshirt-and-blue-jeans-clad, worm- and-soil-obsessed Milwaukeean, Will Allen, is undeniably the world's biggest urban-farming celebrity.

Will Allen, CEO of Growing Power, Inc.

I first learned of Will Allen, founder and CEO of Growing Power, Inc., in a *New York Times Magazine* article titled "Street Farmer," written by Elizabeth Royte in 2009.[1] The fact that a New York–based magazine would spend six pages on an urban farmer from Milwaukee demonstrated Allen's fame in the urban-agriculture world, as well as his rising status in the mainstream. He had received a $500,000 MacArthur Foundation "genius" award in 2008 and had also appeared in other mainstream media such as *O* magazine. He was later named as one of *Time* magazine's "100 Most Influential People in the World" in 2010.[2]

As such, Allen is currently a tireless traveler at the mercy of the urban-agriculture revolution he helped start over twenty years ago. Now in his sixties, the former basketball player turned salesman turned urban farmer seems to be on the road more than he is at home in Milwaukee. When he's not giving his seven-hundred-image slideshow presentation to standing-room-only crowds in the United States and Canada, he is consulting on projects as far away as the United Kingdom, Kenya, and the Ukraine. No stranger to hard work and perseverance, Allen has worked for twenty years to put urban agriculture on the map, and so while the spotlight continues to shine on his six-foot-seven frame, he seems to be on a mission to bring his "Good Food Revolution" to as many people in as many places as he can.

Allen grew up in Rockville, Maryland, and was a typical sports-obsessed boy.[3] But Allen's father, a former sharecropper turned construction worker, also worked a piece of farmland to supplement the family's income and grocery basket. The rules were clear. Allen had to do farm chores, along with maintaining good grades in school, before he could hit the basketball court. Both the academic discipline and the sports paid off. By high school graduation, Allen had his pick of scholarships. He chose to attend the University of Miami, becoming the first African American scholarship athlete there; meeting his wife, Cyndy; and graduating with a degree in education.

A professional basketball career followed, first in Florida, then in Europe. It was while playing for a team in Antwerp, Belgium, that Allen became an unlikely part-time hobby farmer. Allen would spend his off-court time driving around talking to farmers who grew crops in much the same way he had learned from his father. It reawakened something in him and got him toying with the idea of growing food, something he thought he'd never do again. Allen put in a request for a house with a garden—his professional basketball contract included housing arrangements—where he could keep chickens and grow foods like peas, beans, and even peanuts. Allen, Cyndy, and their children moved to the outskirts of Antwerp. He was happy to get his hands into the dirt again, and his teammates and friends didn't mind the free eggs and feasts that Allen would cook with his produce.

Allen was only twenty-eight when he finished his professional basketball career. Cyndy's family had some farmland in Merton, Wisconsin, near Milwaukee, so Allen, Cyndy, and their three children moved there. He farmed and sold his produce at farmers' markets on the weekends, all the while excelling at sales and marketing jobs during the week. Both Royte's article and Growing Power's press materials touch on the discrimination that Allen had to overcome as a black farmer trying to break into the white-dominated farmers' markets in Milwaukee. Tables at the popular Fondy Farmers' Market, the central market in downtown Milwaukee, kept getting passed along to white farmers through private arrangements despite the fact that Allen was at the top of the list for the next available stall through the official process. The market finally agreed to let him set up outside the market, something that never sat well with Allen. He eventually got into the indoor market and was even elected president of the Fondy Farmers' Market, the very group Allen had been shut out of initially.

By 1993, Allen decided he wanted to farm full time and began looking for another site to complement his existing farm in Merton. He found a greenhouse business in foreclosure in an economically depressed working-class part of northwest Milwaukee. The two-acre site was ideal. It had existing greenhouses and was even zoned as agricultural land. As it happened, Allen was able to buy the last remaining farm in the city.

Not long after Allen began operations at this "city farm," neighborhood kids from the largest low-income housing project in the city started coming around, asking questions about how to grow their own vegetables. Allen was thrilled that young people were interested in doing something positive with their time, and so he began to mentor them, setting them up with little gardens of their own. It was at this point that Allen realized he had an opportunity to do something more than grow and sell food with his urban farm. He could begin to address the food-security problems in neighborhoods that lacked access to healthy food. And by excelling as a successful urban farmer, he could create the methods and models for others to empower themselves within their communities in a very positive way.

By 1995, Allen's work with at-risk kids was gaining a bit of a reputation. He was approached by Growing Power, a social enterprise nonprofit group that helped teens acquire work skills. By this time, Allen recognized that he needed some help with his growing enterprise as well. Growing Power, Inc., was formed out of that partnership.

As a working nonprofit and training center with more funding and support staff, Allen found he could really unleash his creativity and test his ideas to maximize food production in a relatively small urban site. Over the next two decades, the original farm on Silver Spring Drive would develop to include a farm produce retail store that would poke out onto the tree-lined residential street otherwise lined with bungalows. Behind it, six greenhouses layered with fifteen thousand trays and hanging pots of herbs, salad mixes, beet greens, arugula, mustard greens, sprouts, and seedlings would become models of high-density planting and an effective example of using every square inch of growing space. The gurgling water sounds in each greenhouse would come from Growing Power's innovative "aquaponics" system, a closed-loop nutrient cycle where the tilapia and perch created fertilizer for the food plants, which would then filter and clean the water for the fish tanks. Each greenhouse would also have worm bins—lots and lots of worm bins—each worm eating its weight in compost and turning it into rich, black soil for the plants. Two large aquaponic hoop houses would contain more fish and growing beds for salad mixes and seedlings. Seven more hoop houses would be used to grow salad greens and mushrooms year-round, thanks to Allen's ingenious idea of mounding compost up the sides of the polyethylene plastic sheeting stretching over the hoops, thereby creating a heat source to keep the hoop houses producing, even throughout the well-below-freezing Milwaukee winter season. Growing Power's farm would also be home to fourteen beehives as well as three poultry houses with laying hens and ducks. Goats and turkeys would be the resident livestock. Allen would innovate and refine urban-agricultural systems like large-scale composting of food waste that would otherwise go into Milwaukee's landfill. The site would be Allen's living laboratory for the model of intensive, mixed, modern farming on an urban lot that was becoming increasingly famous.

By the time of Elizabeth Royte's visit, Allen's "agricultural Mumbai, a supercity of upward-thrusting tendrils and duct-taped infrastructure," as she described it in her article "Street Farmer," the two-acre Growing Power flagship farm and Community Food Center on Silver Spring Drive in residential Milwaukee, was producing fresh food for ten thousand of the area residents, generating $250,000 in annual sales, and training legions of young, eager urban farmers and community food-security activists.[4]

I caught up with Allen when he came to Vancouver, Canada, for a Simon Fraser University–sponsored talk in January 2011.[5] Over eight hundred people crammed into the community league hall, almost at maximum capacity, despite the wet, cold winter evening outside. There were people of all ages, from toddlers to grandparents. But the largest age group, overwhelmingly, was made up of people in their twenties and thirties.

Will Allen, founder and CEO of Growing Power, Inc., holding a fish from the aquaponic tanks. Photo © Growing Power, Inc. Used with permission.

Dressed, as always, in a worn baseball cap, blue hoodie, and blue jeans, Allen climbed to the podium slowly—he was recovering from knee replacement surgery a month earlier—to a standing ovation. Likely sensing the near-religious fervor for urban agriculture in the room that night, he prefaced his talk with a plea that "people shouldn't get so hung up on this idea of 'urban' that we forget about our rural farms." He urged the crowd to do their best to support local rural farmers. "They're under a lot of stress. They are just as important." Even though he was there to talk about his Growing Power projects, essentially his successful approaches to urban agriculture, he wanted to make it clear that his "Good Food Revolution" was all about local—both rural and urban—at the end of the day. It was about a partnership where good, healthy, accessible food was being produced by both farmers in the city and in the surrounding countryside.

After his appeal for respect and support for rural farmers, Allen fired up his slideshow presentation. He began with a slide from the early 1990s of four preteens posing on piles of compost, goofing off for the camera the way kids do. He then asked the audience what the difference was between the kids seen in the slide and kids today.

"They're not overweight," someone near the front shouted.

"They don't have cell phones," was another idea.

There were a few other unremarkable suggestions, until Allen finally said, "No, their pants are pulled up."

It may be a joke that Allen tells at every one of his presentations, but the delivery was genuine and the joke diffused a bit of the cloying earnestness in the community hall that evening. "Some of these kids are thirty years old now," Allen added quietly, as if for his own benefit, just before he moved on to the next slide.

Allen showed another slide of the very early days of the urban farm. Obviously a mixed farm in the city drew some attention from the neighbors, who originally were opposed to goats, chickens, and a farm stand but were easily won over with fresh eggs and salad greens, Allen said. He also started to work with kids in the neighborhood who would otherwise be getting into trouble, or who had already been in trouble. Together, they established little flower gardens and grew vegetables that the kids would give back to their community. Kids who were working off minor criminal offenses could do so by growing food on community land and donating the food back to the community.

Allen clicked between a "before" slide of a street scene in a crime-ridden part of Milwaukee and an "after" slide of one of his "flower explosions." He explained that he found early on that the mere act of blanketing a streetside boulevard with flowers would remarkably reduce smash-and-grab break-ins in parked cars. Drug dealers also had an aversion to flowers in their favorite parks where they dealt. Allen laughed at the thought of it but came to call his flower explosions and food gardens "effective crime-fighting tools."

Allen also figured out that he could grow soil just by diverting compostable waste from Milwaukee's landfill. "Everybody has seen those big green trucks rumbling through their neighborhood," said Allen, as he showed a slide of a large garbage dump truck. "That is our competition." No matter what the weather, Allen wants Growing Power to get to that valuable green waste before the waste companies do. Allen needs as much soil as he can create if he wants to keep growing and expanding. "We picked up twenty thousand pounds of waste from just one supplier last year," Allen proudly declared, showing a photo of a pile of unsold fruits and vegetables that never made it out of their packing crates. Cardboard and all got piled up in compost rows. "Then we feed it to livestock," Allen cheekily offered before advancing the slide. "These are our livestock." Allen smiled, finally showing a slide of a wriggling mass of red worms. "We have billions of employees, and they work for food only and don't talk back," he laughed. "This makes us the largest employer in the world."

The worms, Allen's passion, eat their weight daily in waste, he repeated over and over. Growing Power, or, more to the point, the red wigglers in Growing Power's employ, compost over one million pounds of waste in Milwaukee, creating precious nutrient-rich soil rather than greenhouse gases in the Milwaukee landfill. Allen was originally told that those worms wouldn't survive the northern Midwest winters, but he found a way around that, too, by overwintering them in huge compost hedgerows large enough so that the worms could dive toward the center of the piles where the biological activity of the decomposing compost kept them from freezing. "We started with thirty pounds of worms, now we have five thousand pounds." It's the cornerstone to his intensive crop production. "You can drop me off anywhere in the world with a handful of worms, and I can create a growing system." That is the power these "employees" possess with their digestive capacities. "If you remember one thing tonight," Allen repeated to the crowd one more time, "this is all about the soil. Without high-nutrient soil, you cannot grow healthy food. The taste of the food is due to the richness of the soil."

Besides worms, Growing Power now produces one hundred thousand fish—tilapia, perch, blue gill, and a few blue koi—in its aquaculture tanks (the fish waste becomes fertilizer for plant crops, which clean the water in a convenient nutrient cycle). There are thirty-nine goats for the artisanal cheese production and honeybees that produce honey sold under a house label called Urban Honey. There are five hundred laying hens producing eggs. There are even some heirloom turkeys that Allen is trying out as a personal project. And lots and lots of green crops, year-round, thanks to the greenhouses and hoop houses outside of which Allen piles compost that creates heat to mitigate against the bitter Midwest winters, and that allows almost year-round production of salad crops. And no growing space is ignored. Baskets hang from the greenhouse frame, and trays of sprouts and seedlings are layered to make every cubic inch count in a space about the size of a small supermarket.

But Growing Power's ambitions far exceed even what has been accomplished in Milwaukee. Its mission statement asserts that it is "a national nonprofit organization and land trust supporting people from diverse backgrounds, and the environments in which they live, by helping to provide equal access to healthy, high-quality, safe, and affordable food for people in all communities."[6]

There are currently fifteen sites under the Growing Power umbrella, and spaces are constantly being absorbed into its orbit. Allen clicked to a slide of an unused greenhouse he recently took over in a cemetery in Milwaukee for growing space. "Cemeteries used to grow their own flowers. Now we keep people away from the great boardroom in the sky a little bit longer."

For the past decade, Erika Allen, Will's daughter, has been managing

Growing Power, Inc., layered aquaponic and plant nutrient loop in motion. Photo © Growing Power, Inc. Used with permission.

and expanding Growing Power sites in Chicago. In 2002, the Fourth Presbyterian Church in Chicago bought an unused basketball court and partnered with Growing Power to transform it into a community garden. It is now a community farm that gives residents in this community access to organic vegetables, nutritional education, and work-skills education. In 2005, Growing Power created a 20,000-square-foot (1,858-square-meter) urban garden/space that was established as the Grant Park "Art of the Farm" Urban Agriculture Potager. This raised-bed vegetable garden contains 150 heirloom edible flowers, herbs, and vegetables in the heart of the downtown region. Two years later, Jackson Park Urban Farm and Community Allotment Garden was set up to provide residents of Chicago's South Side with food-growing space and training with Growing Power volunteers. In 2010, Growing Power started a 2.5-acre (1-hectare) urban farm called the Altgeld Gardens Urban Farm on the city's South Side in one of the city's most disadvantaged communities. In its first year, Altgeld Gardens employed 150 adults and 40 at-risk youths, as it grew and distributed healthy fresh food to the area residents. "Growing Power is one of the only multicultural, multigenerational organizations led by a person of color in the US," Allen said.

Growing Power, Allen explained, is now a major training center for urban agriculturists and an innovation hub for cutting-edge urban-food production, renewable-energy projects, and waste-reduction systems. In 2010, Growing Power welcomed 15,000 visitors to its sites, engaged 3,500 volunteers, graduated 32 commercial urban-agriculture farmers, and 1,000 future urban farmers got a taste of what lay ahead for them there.

Growing Power is also the economic model for other successful nonprofit social enterprises, with its diverse, multilayered revenue streams. These include revenue from educational and training programs; the sales of foods to restaurants and cafeterias through the retail store, at farmers' markets, and via CSAs; foundation grants; and Will Allen's near-constant speaking engagements. In 2010, Growing Power had a $4.1 million operating budget and generated a $700,000 surplus.

Allen explained that his next challenge is to figure out how to incorporate renewable energy into urban agriculture. He is experimenting with solar panels. "Renewable energy must be part of the energy cost to run this high-value space."

Lastly, Allen unveiled the architectural renderings for the next big project, a five-story vertical farm that will serve as the new growing, research and development, training, and administrative headquarters in

Milwaukee. (On December 31, 2010, Allen held a press conference in Milwaukee to announce that Growing Power's 27,000-square-foot [2,508-square meter], five-story vertical farm's building permit had been approved by the city. Construction was scheduled to start in 2011.)

The impressive financials and the high-tech vertical farm seem to prove a point about the economic viability of urban agriculture to those who can see value only in dollar signs and ambitious building projects. More impressively, Growing Power has revitalized a city's food system and created a locally controlled, sustainable, green economy that provides good, healthy food to people who otherwise don't have the opportunity to access it. It provides training to those whose options are limited. And it builds and improves communities where other social programs have failed.

"Food is a very powerful organizing tool," concluded Allen. More to the point, Allen has proven that you can't really conceive of community-development plans anymore and leave food off the map. It has to be the central organizing element, and other issues, like literacy, crime reduction, urban revitalization, improvements in physical and mental health, will naturally follow. And even Allen admits that the tide has finally changed; people are thinking of community-based food systems as viable and important to the urban landscape when, "in the beginning, a lot of this was just talk."

He even has interest from some surprising groups. The slide showing a group of buttoned-down, middle-aged, white Walmart executives looking intently at tilapia tanks got a raucous laugh from the leftist Vancouver crowd. "You may laugh," Allen countered, "but one of the things about urban agriculture is that we need everybody at the Good Food Revolution table. Those days are over when we exclude people and organizations from being partners." Allen was clearly focused on big changes, not little ones. And the challenge ahead is clear. Industrial agriculture, and the systems that exclude some people from being able to access good, nutritious foods, isn't going anywhere soon. "We need fifty million new farmers" to make a permanent, meaningful change in how we produce, access, and recycle our food.

The success of Growing Power is undeniable. There can be no doubt that Allen is the right person at the right place at the right time. He's worldly but rooted in his community. He's a philosopher and a farmer. He's overcome many obstacles but exudes pure positivity and hope. He's equally at home talking to halls packed with people as he is on the farm. (The fact that he still operates his own farm gives him dirt-under-the-fingernails credentials in the agricultural world as well.) As a role model, Allen is a very successful African American farmer (and CEO), which is an important leap forward in a food-security movement in the United States where racial polarities seemed unbridgeable at times. Within the African American landscape, Allen is a positive example of a successful farmer in a world where the history of labor abuses of sharecropping often gnaws at the black community. But Allen acknowledges this fact head-on and encourages people to see the healing that food growing can accomplish in communities of color. He's making it a desirable and viable way for communities to find a way out of poverty and poor nutrition among families who find their options few and far between.

And though Allen seemed weary after his whirlwind visit to Vancouver, with the constant rat-a-tat questions and people merely wanting to shake hands, he graciously made time and personal connections until the last hand was grasped. And then he was off to another talk, another expert panel, in Atlantic City, because, for the time being, Will Allen's Good Food Revolution rarely affords him a rest.

Chapter 12

DETROIT

Praying for an Economic Revolution

From Motown to Hoetown
—Headline in *Toronto Star* newspaper on
Detroit's move toward an urban-agriculture
revolution, September 26, 2009

When Detroit was originally settled in the eighteenth century, it was an arrangement of long but narrow strip farms running like piano keys along the riverfront so that a maximum number of farmers could have access to the waterway. They were called "ribbon farms," and many street names in the city still carry the names of the original farmers. Detroit, of course, is best known as the great industrial womb of the American automobile industry that brought jobs and prosperity to the city for the first half of the twentieth century, and that launched the car culture, whose effect we continue to see on everything from our expanding waistlines to our car-centric urban planning in every North American city. Despite the "Motor City's" legacy, the last sixty years have been disastrously unkind to Detroit. The situation at this point is beyond desperate. It is truly a city in need of a miracle.

At its peak prosperity in 1950, Detroit was home to almost two million residents. Fifty years later, the US Census Bureau counted about 900,000. These days, the number hovers around 800,000.[1] The current thinking is that the population will level out at 700,000.[2] It is no secret that the Motor City is now rather empty of cars, as it is of homes, people, and food. And with the ongoing economic recession in the United States, there really isn't anything on the horizon that would suggest that Detroit has even hit the bottom of its decline. The city is now typically held up as the poster child for a shrinking postindustrial municipality on the brink of bankruptcy, with crumbling civic infrastructure and a city-wide food desert for the citizens who still live there. Where else can you find a city without one single national chain grocery or big-box outlet?[3]

When Detroit makes news, it's usually due to its unemployment rates, political scandals, crime rates, and economic decline. In 2005, Detroit was the "most impoverished city in the nation," according to the US Census Bureau.[4] The unemployment rate officially hovers around 30 percent,[5] but many think it's as high as 50 percent. Thanks to unemployment and a vanishing population base, the city's treasury has a deficit of $300 million, so it can't provide even the most basic services, such as police and firefighters.[6] And its education system is in shambles.

A succession of inept mayors has only made things worse. Homes are virtually worthless in Detroit. The median price of a home that a person can actually sell in Detroit is $7,500,[7] yet I was told repeatedly when I was there that you could easily pick up a house for less than $1,000. In fact, the usual comparison is that you can buy a home in the city of Detroit—not the tony suburbs, which, strangely enough, sit just on Detroit's periphery—for less than the yearly cost of insuring a personal vehicle.

As the saying goes about desperate times leading to desperate measures, Detroit is at the point where all ideas for urban renewal are being considered. Detroit sprawls over 139 square miles (357 square kilometers) just in the city proper.[8] Its footprint is equal to that of Manhattan, Boston, and San Francisco combined. Interestingly, it is the urban-agriculture groups—both the social justice community gardens and co-ops and the hardcore capitalists—that have stepped up to the plate to take a swing at what might be the daunting task of stemming the exodus of residents and making Detroit a livable community once again.

Food-justice advocate Alethia Carr grew up in Detroit in the 1960s. Speaking with me over the phone—a rather enthusiastic winter snowstorm in Detroit derailed our scheduled face-to-face meeting—Carr told me that she had a peach tree, a cherry tree, and a crabapple tree right in her backyard at her childhood home on the east side of Detroit.[9] There were pear trees and even grapes growing in neighbors' yards, too, she remembered. "I was told that the area was originally an orchard," Carr recounted.

I had contacted Carr because of her work surrounding food-security issues in her hometown. She is both a fellow of the Institute of Agriculture and Trade Policy (an international nonprofit food-security think tank) and a longtime public health director in charge of maternal and child health programs in Michigan. With both a personal and a professional interest in food security in her community, she has had a hand in many Detroit grassroots initiatives related to nutrition, health, and issues surrounding access to healthy and affordable foods in urban areas. I was hoping that she could help me understand the food-security issues there. But for the first ten minutes or so, we ended up swapping food memories of our respective childhoods, and the similarities were surprising.

We both remember lively, productive backyard food gardens easily found behind fences. We both grew up in neighborhoods that had at least one small, independent grocery store within walking distance that actually stocked produce, vegetables, flour, sugar, butter, and other staples, and had a fresh meat counter at the very back. Fresh, seasonal foods were the stuff of homemade meals (which were the norm in both households), and life happened out on the streets, not inside the house, as it does nowadays. So what happened between then and now, I wanted to know, that we could have had such similar childhood memories despite our slight generational gap, as I looked out over a city with empty streets, derelict residential block after derelict residential block, and boarded-up commercial buildings right in the downtown core?

Carr walked me through a more personalized history between the

1960s and the Detroit of today. Carr, an African American whose father worked in the automobile industry, remembers a "pretty well-integrated" city as a child. Where Carr grew up, it was a predominantly African American part of town. And it had those corner grocery stores "for a period," she said.

Detroit's economy started its decline as automation replaced the need for humans in manufacturing. Unemployment rose, and social tensions stretched to the breaking point, fueled by abuses of the white police power that ran unchecked against black Detroiters. (A Saturday-night police raid on a black nightclub in July 1967 finally touched off the powder keg that became the infamous five days of chaos and violence known as the Detroit Riots.) "White flight" to the suburbs, which had already begun due to jobs leaving Detroit as early as the 1950s, accelerated. Over the next few decades, Detroit suffered the slow, steady decline of a dying industrial city.

As industry and jobs left, the city's tax base shrank, making it less desirable to the businesses and residents it needed to attract. The suburbs sucked out much of the middle- and upper-wage earners, hollowing out the city. Grocery stores followed the migration to the suburbs. Those who still lived in the city gladly drove out of the city to the big-box stores to shop anyway.

"My in-laws had a grocery store," Carr said, as we talked about the consolidation of the major chains and how they seemed to be preferentially located in new suburbs. "It was a small mom-and-pop, but they tried to carry the same brands as the big stores. But their orders were too small." Soon grocery distributors dropped these smaller independents from their supply chains. Independents had to buy at retail or public wholesale and become resellers. Eventually the costs became too high for the customer base.

Those neighborhood grocery stores either vanished or changed their inventory. "Grocery stores went from carrying a wide variety of groceries, to carrying a wide variety of beer and liquor," Carr lamented.

It is a commonly repeated fact that there are no national chain grocery stores left in the city.

Carr's professional career gave her another unique perspective on food security in Detroit, above and beyond a lifetime of living there. She has been involved in public health for more than twenty years, after completing her postsecondary studies as a registered dietician and her master's of business administration. Not only is Carr the maternal and child health director of Community Health for Michigan's Department of Community Health, she served as the director of Michigan's Women, Infants, and Children (WIC) program for nine years. Michigan's WIC program is the eighth-largest in the country and serves 230,000 clients each year.

The Special Supplemental Nutrition Program for Women, Infants, and Children is a nationwide program to "safeguard the health of low-income women, infants, and children up to age five who are at nutritional risk by providing nutritious foods to supplement diets, information on healthy eating, and referrals to health care."[10] Women and children are usually the most affected by food insecurity, so in 1974, the US government established the WIC program for those individuals and families whose low income affects their food choices. There are currently 9.17 million women, infants, and children who benefit from the WIC program every month in the United States.[11] "Michigan's program has a debit card loaded with specific food choices," explained Carr. Those choices are healthy, nutritionally intact, whole foods.

"The challenge we see in many young families is how to cook, or how to cook with whole foods, not those boxes filled with things that are high in sodium and high in sugar." Young families, says Carr, simply lack the knowledge of basic cooking skills and how to properly balance their meals. But access to good, fresh food is still a big concern.

Urban gardening became an effective tool to combat this "food-literacy gap" as well as the lack of access to fresh, healthy food, especially in low-income areas. The lack of equal access in Detroit's black community is of particular concern because this community suffers the most food insecurity in relation to other cultural and ethnic groups. Urban agriculture has been looked at more seriously in the last couple years as a way to remedy this discrepancy. "Our local health department has been concerned with poor access to food here in the city and what can be done. The planning commission in the city has had urban agriculture on their mind for the past eighteen months."

As a result of both a grassroots push and encouragement from social and civic programs, urban agriculture has taken off in Detroit. Unlike in many cities, vacant land is one thing that Detroit has in abundance. With over 30,000 acres (12,140 hectares), or 40 square miles (103 square kilometers), of vacant land in the city, the potentially available land is larger than the entire footprint of Miami.

One international nonprofit organization, Urban Farming, founded by musician Taja Sevelle, claims to have created five hundred urban gardens so far in Detroit where all the produce grown is free for the community.[12] The Greening of Detroit, another nonprofit founded in 1989 primarily as an urban-reforestation project, now supports over two hundred school, family, and community gardens, and is seeing a 20 percent growth in urban-agriculture projects year after year.[13] The Detroit Black Community Food Security Network expanded its two-acre D-Town Farm to five acres in 2011.[14] By 2010, Detroit had 1,300 urban food gardens scattered among different groups.[15] These projects are making important inroads to addressing the accessibility of good food and to empowering Detroiters.

But Detroit would need tens of thousands of these smaller initiatives to start to tip the scales away from urban blight and to start to address Detroit's unemployment crisis. I finally told Carr that the main reason I had come to Detroit was the headline-grabbing announcement in 2009 that got not only my attention but that of the major news media in the United States. And I wanted Carr's opinion on a controversial, ambitious initiative that has been proposed in Detroit: the world's largest private urban farm.

HANTZ FARMS

John Hantz is not a farmer in search of land. He's a Harvard-educated, self-made stockbroker, financier, and businessman—his personal fortune is estimated at $100 million—in search of a solution to a crisis in his hometown of Detroit, Michigan. Hantz moved to Detroit over twenty years ago to turn around the then-floundering Detroit office of American Express financial services, which he did, making it one of the top performers in the country. In 1998, Hantz left American Express and has since built nothing short of his own empire. The Hantz Group is a diversified business empire of five hundred employees that includes a private-jet-charter company (Hantz Air), a money-management company with over $2 billion under management (Hantz Financial Services), a beverage company (United Beverage Group), and a lumber liquidator.

Until 2009, however, most Detroit residents hadn't even heard about John Hantz when newspapers and magazines picked up the press release announcing the planned Hantz Farms project. The initial press release issued by the Hantz Group on March 23, 2009, stated that "preliminary plans for a newly developed urban farm within the City of Detroit will utilize vacant lands and abandoned property to create Hantz Farms, the world's largest urban farm."[16] The first phase, the press release continued, would "utilize more than 70 acres of underutilized vacant lands and abandoned properties on Detroit's lower east side." Once the city approved the plan, "the farm would be operating within six months." The bold plan was not well received.

The initial reaction was suspicion and alarm at a potential landgrab, especially among those who had been working in the city's food-security community. "Huge alarms, like sirens going off," was how Carr said the news of the Hantz Farms project was greeted by many in Detroit's food-security networks. The concerns were that this farm would happen at the expense of the city and that there was no stated intention to share the wealth that might be created with the city residents. Most importantly, Hantz was accused of taking advantage of the desperate situation in the city.

Let's not forget, John Hantz is a white multimillionaire in a city where 82 percent of the population is African American. Fortune often follows racial lines in Detroit, a city with chronic unemployment that sits at around 33 percent and social problems heaped upon poverty issues like crime, drugs, and lawlessness. But Detroit's biggest problem is too much land and too few people. The municipal government has a budgetary deficit of $300 million. The population is in a steady decline, with thousands of people leaving every month—an average of sixty-five people per day, according to 2011 statistics.[17]

URBAN FARMING SUMMIT: THE BUSINESS OF URBAN AGRICULTURE

John Hantz, self-made millionaire and one of the few successful businesspeople who still live in the city of Detroit, has a different vision. In April of 2010, Hantz spoke publicly and personally about his urban farm idea as part of the panel discussion "Urban Farming Summit: The Business of Urban Agriculture" at the University of Michigan–Dearborn.[18] He wanted to explain his rationale behind the idea for the private farm, knowing that his critics would be in the audience, many of whom were openly suspicious of his motives for creating a huge private urban farm. Many hoped that Detroit would adopt a more cooperative, socialist model of common land ownership and common profit sharing. The city government was going to own large parts of the city by foreclosure anyway, and the idea of urban-farming cooperatives leasing cheap land from the city—a model based on Cuba's successful urban organic farming program—had also been put forth as a way forward toward urban revitalization.

For twenty years, Hantz has driven from his house in Indian Village (a cluster of around 350 century-old homes on Detroit's east side surrounded by urban decay) to his office in Southfield (a clean, orderly, pleasant-looking suburb of Detroit). "Every year I tell myself 'It's going to get better,' and every year it doesn't," Hantz told the crowd of businesspeople, academics, community activists, and citizens in Detroit's nonprofit sector who came out to hear from the person behind this radical and controversial plan for Detroit's revitalization: the world's largest urban farm.[19]

One day in 2008, Hantz explained, on his usual drive through derelict residential and commercial major streets, his optimism turned to pessimism. He realized that nothing was going to happen and that Detroit would just continue its long slide into decay and ruin. Someone needs to do *something*, he realized, not wanting to accept the dire scenario. "And then I got that hated, dreadful feeling, that that someone might be me," Hantz said, sending a chuckle throughout the room, breaking the tension somewhat. The fact that he didn't know what the solution was, however, bothered him.

He thought about what the city needed to bring people back, to get them to invest again in Detroit, especially given the well-documented negatives that hang over the city's image. "We have a crime issue. We have a school issue. And we have a [tax-base] issue." How exactly do you attract people to invest their money and lives in a city that is still spiraling downward?

Hantz looked at the problem as he would any other investment situation. How do you remedy the root problem in Detroit—that land no longer has any value? What would be a positive solution to this problem that would enhance land values and create opportunities, and that people would actually want to live around? What would make the land valuable enough again to attract people and businesses back to Detroit?

In the business world, Hantz explained, value is created through scarcity. Land had to be taken out of circulation—a lot of land, he figured—to create scarcity. Otherwise no one would invest in Detroit because the value of land would just keep dropping. "If I was in Chicago, and I saw an open lot next door to me that went up for sale, I'd move to buy it," Hantz used as an example. "Because three other people might also want it. In the city [of Detroit], if a lot becomes open, you wait. Because next year it might be cheaper," Hantz said.

But Hantz knew that his critics were saying that the city shouldn't jump at the first offer of a deep-pocketed businessperson looking for cheap land, fearing it was an out-and-out landgrab. Many groups had expressed the idea that the large amount of available land is Detroit's biggest asset—that perhaps it's the city's only remaining asset.

"There's a difference between green space and vacant land," Hantz said, preempting the inevitable argument on this subject. "Green space is a planned process that you elect to do as citizens to improve quality of life. Vacant land is a train wreck. It destroys value. It destroys communities when it doesn't have a purpose." And, Hantz argued, it is a fact that vacant land's truly destructive side is that it consumes the city's resources.

Hantz put forward the fact that a vacant parcel of land in a city has a carrying cost of about $12,000 over a five-year period. Detroit has those two hundred thousand vacant parcels (most through foreclosure) scattered around the city, draining money, requiring labor to maintain, and using city services like police and fire protection. (Executive of planning for the City of Detroit, Al Fields, who was also on the discussion panel, confirmed that this figure was accurate. He then added that it is a cost that in many cases prevents the city from carrying out its duty to provide those essential services because of lack of funds.) "And we wonder why we have a $300 million deficit," Hantz rebuked. By Hantz's calculation, all this vacant land—seen by some as an asset for the city—comes with a $3 billion carry cost over five years. What would you do with an "asset" like that? Hantz questioned. "You'd pawn it off on somebody. As quickly as possible!"

So that "dreadful feeling of not knowing what to do" gave way to a revelation. A large-scale farm, right in the city, would begin to take large amounts of land out of circulation "in a positive way." Productive land that is taken care of and maintained might be something that people would want to live around, as opposed to homes left to crumble or burn at the whim of the rampant arson attacks that take down many vacant homes. (Demolishing a home in Detroit costs over double what the city could recoup by selling it. This explains why so many foreclosed and abandoned homes are left to crumble or to be burned to the ground by arsonists.) Detroit, Hantz reminded the group, has over thirty thousand acres of vacant land, and he noted that Detroiters "act like we have five." Hantz decided it was worth a shot. He committed $30 million of his personal money toward an experiment in what it might actually take for a place like Detroit to turn the corner: Hantz Farms.

Perhaps the term *farm* was a bit problematic, as was the initial ambitious scope of buying up seventy acres all at once. Some imagined tractors, combines, and large industrially farmed monocrops. Others were trying to work out how a traditional mixed farm, writ-large, perhaps with livestock and all, would look in an urban environment. Hantz, on the other hand, with no farming background, looked at the "farm" concept in a new light.

Hantz Farms would incorporate hardwood forests, Christmas trees, and even morel mushroom fields. It would reuse old commercial structures to experiment with indoor, year-round growing systems. It would basically throw the proverbial kitchen sink at the farm, because no one knew what the right mix was going to be. And Hantz wasn't wed to one idea or another. In fact, he wasn't even dead set on the farm itself. It was just a means to an end, to bring value back to land in Detroit, to get people investing in building communities again, and to make Detroit a city where Hantz didn't have to live next to vacant lots.

"I would trade in Hantz Farms tomorrow if all that I could get from you was to agree that we need to homestead all the acreage." Hantz feels that a Detroit Homestead Act would be a much better solution than his farm because it would establish the same outcome, taking all that land out of circulation and therefore making land valuable once again. And there would be tens of thousands of entrepreneurs coming up with ways to make their land valuable, not just one in charge of a lot of land. "Either way, I don't have a lot next to me that's vacant!"[20]

Hantz then quoted statistics from the Homestead Act, the federal program in the United States that put 270,000 acres, 10 percent of the US land base, into agricultural use under the ownership of two million citizens between 1862 and 1934. It's a concept that has experienced a revival in certain dwindling towns and rural communities where land will be given to individuals willing to homestead it.

(Beatrice, Nebraska, has enacted its recent Homestead Act of 2010.[21] People can apply for land by downloading the online application. If the application is approved, the regulations state that a nine-hundred-square-foot (minimum) home must be built on the land, in which the owner must live for a minimum of three years in order to secure title to the land. The fee for the Homestead title is simply the cost to file the paperwork and cover land transfer costs. Dayton, Ohio, and Grafton, Illinois, have also jumped on the recent Homestead Act revival wagon.)

For-profit agriculture would also be a major employer, Hantz argued. Carlton Flakes, a social worker who manages an urban-agriculture training program for ex-offenders transitioning back into society had already spoken to the need for jobs as one of the panelists. "I keep hearing about the thirty thousand vacant acres. I've got thirty thousand ex-offenders, most of whom need viable employment. We have a workforce that's underutilized, and believe it or not, willing and anxious to be productive and to work."[22]

Near the end of the one-and-a-half-hour discussion, the question that clearly was on the minds of many in the audience was finally raised. Many of the people wanted to know: Isn't this just a landgrab?

"It is definitely a landgrab," Hantz shot back. "You can't farm without land." He then explained his belief that he'll do a better job, faster and with better efficiencies, if at the end of the day he's improving *his* land value, if there's something in it for *him*. That's how capitalism works. And the only way to get people like him to take the risks involved is if there's a reward. "We've already proven that capital is a major part of our problem," continued Hantz. "And capital gets tired of being insulted, just like nonprofit gets tired of being insulted." Opportunity for profit simply has to be a part of the equation if people are going to invest.

"You could do every idea I've heard today, ten times over, right now," stressed Hantz. He pleaded that this was Detroit's time to take some risks, to learn, to make mistakes and improve from those mistakes. "We're not in a perfect situation right now," he concluded. If entrepreneurial spirit is allowed to flourish and isn't hampered as it currently is in Detroit, Hantz sees a day when the city is the "go-to place for urban agriculture systems," a type of global showroom for every conceivable scenario that any community around the globe might want to implement. Entrepreneurs and governments would come to Detroit, tap into the expertise, look at the models, and buy the technology developed in Detroit.

MIKE SCORE, PRESIDENT OF HANTZ FARMS

My cab driver didn't have much trouble finding the headquarters of Hantz Farms, a modest one-story brick building. Occupied buildings stand out like a sore thumb in southeast Detroit, especially along Mount Elliot Street, which had obviously been home to a thriving commercial strip of businesses at some point in the past. Now, yellow bags announcing foreclosure notices are stapled to most doors where buildings are still standing. The cab driver refused to let me out, even though we found the building and it looked occupied. "No, this is a bad part of town," he warned, despite the street being devoid of people or other cars. He insisted that I phone and ask someone to meet me at the cab, or at least at the door.

Mike Score, president of Hantz Farms, came to greet me, looking every bit the farmer, with his cowboy boots under dark pants, suspenders over the work shirt, and a neatly trimmed beard. He could pass for an Amish farmer given the right hat. But instead, he's a city kid. "I was born in a house two miles from here," Score told me as a fire crackled away in his office fireplace on the bitter February 2011 day when I visited.[23]

Score spent his early childhood in southeast Detroit, his adolescence and teen years in a nearby suburb, and only came to agriculture because he hated his business major in the university and switched into a degree in crop and soil science after just one semester. His alma mater, Michigan State University, hired him fresh after graduating, and Score started working as an agriculture extension agent—basically a consultant who helped farmers tweak their businesses for efficiencies and profitability.

Score built his career by helping fledgling food and agriculture companies write business plans to scale up, change course, and generally shift with the winds of agriculture. When John Hantz approached the University of Michigan to find a person to write a three-year business plan for an urban farm, Score was the consultant they suggested.

It was a dream assignment, Score confided. He'd already been fantasizing about bringing agriculture back into Detroit. It took him a mere three days to submit his detailed plans and financial projections.

Hantz was impressed and asked for a longer six-year projection. Again, Score delivered. Little did Score know that this had been a job interview, John Hantz-style. In December of 2009, Score was announced as president of Hantz Farms, and a major part of his job in the past two years has been to "change the dialogue."

"There'd been some ugly rhetoric," said Score. "A big white farm in an African American city . . . it sounds familiar, you know." But Score had been meeting with and talking to critics of the plan, listening to concerns and answering questions. He went door-to-door in the neighborhoods, asking the people who were left how they'd feel about having a farm in their community. According to Score, the unanimous response was a resounding "yes, please."

The frustration now is familiar for anyone who understands civic politics.

I pressed Score for a launch date of the project that had created so much buzz in urban-agriculture circles around the world. "We used to guess," he eventually replied. "We're three years into this now—over one million dollars of development costs for the project—and we're just now getting to the point where we might be able to buy 1.8 acres from the city's inventory." That is a mere drop in the forty square miles of inventory (and counting) of foreclosures that had been absorbed back into the city land bank.

"If there's a filter to help keep bad projects from happening in the city of Detroit, then I would say that it's a very effective filter. Probably too effective," laughed Score. I imagined, however, that humor was just one reaction to the glacial pace at which most cities, even the ones that run efficiently, operate.

I suspected that Score is an optimist. Most people in agriculture are, so I kept pushing for even a wild guess at planting. "The best-case scenario is that we'd get approval on the first block of land this spring. By fall of 2011, a substantial part of the site prep and the first perennial crops would be in the ground."

The initial project proposed by Hantz Farms was seventy acres. It was too big and too much for the city to digest all at once, Score conceded. It has since been split into bite-sized parcels that will be implemented in phases. The first trial farm site is planned for the immediate area surrounding Hantz Farms' office, if it can get the go-ahead from city hall.

Behind a desk, taped to the wall in his office, Score had three surveyor's maps of the areas surrounding the office showing property lines and legal descriptions, about twenty acres in total. There were green dots indicating properties that Hantz Farms had already bought, amounting to about one acre, but the dots were piecemeal at this point. I noted that the yellow dots were the majority. "The yellow dots are city-owned," explained Score. The rest of the parcels were privately owned, but Score's scribbles on almost every one indicated how far behind they were in taxes. "There are only a few houses in this whole area where people are current on their property taxes. Everyone else is two-to-three years behind and heading for foreclosure." Despite the twenty acres immediately surrounding the Hantz Farms head office that will come under the ownership of the city in the next few years, the scaled-down first phase of Hantz Farms will be a five-acre test site.

We had been discussing plans and roadblocks for over an hour in the office when Score proposed "a field trip." He suggested a drive through those neighborhoods we'd been discussing, so I could see the landscape for myself.

Score narrated nonstop as we crept along the streets in his Ford® Focus. The houses that were still standing—"It costs $10,000 to demolish a house"—had roofs that had caved in, front doors swung open, and furniture strewn on front porches. "The median house price in Detroit is $8,000 right now," Score said, echoing the figure of around $7,500 as an average house price that I kept hearing over and over again while in Detroit. But in this part of town, houses that sell usually do so for less than $1,500. Yet "it costs the city nine million dollars per square mile, that's $360 million in Detroit per year, just to provide basic city services." That's why the city's deficit just keeps ballooning.

As we drove up and down the streets, a mishmash of residential teeing-off of main roads where abandoned commercial properties now sit, I commented that there were basically no people left to displace. "We won't evict anybody from their property," Score said and then pointed out a single home that was maintained. "Mattie can stay if she wants to." But in truth, there aren't too many people to work around. In the ten-acre area that we drove through, Score remarked that only two houses were occupied.

Houses that hadn't been burned by random acts of arson or during the annual Devil's Night frenzy the night before Halloween have been left for looters to destroy and for time and elements to do the rest. We passed by houses with their roofs caved in. There were streets that were completely abandoned. One house was only half torn down; a toilet and sink were still intact on the second floor like a cross-section had been taken from the house. Empty lots had become dump sites. It was the kind of landscape you normally associate with either war zones or disaster areas.

Score's optimism was remarkable. I couldn't get past the half-standing and burned-out homes. "We'd like to plant Christmas trees here for a U-pick," he cheerfully remarked as we drove through a neighborhood that I wouldn't want to find myself in past dark. The soils in residential areas, Score continued, are actually in good shape because they've been covered in grass and have not been exposed to industrial

Just one of several dozens of burned-out and destroyed houses in a residential neighborhood near the Hantz Farms head office, Detroit. Photo by author, February 1, 2011.

waste. Soil remediation will be easy; fruit trees and vegetable gardens can be planted. In some cases, raised beds with barriers between the existing contaminated soils and the new growing soil might be needed.

The former industrial and commercial land is a different story, but there's a plan for that, too. Score says that because of all the concrete and plaster used in the city over its history, the soils in certain areas are typically quite alkaline now, with a pH of around eight. "It changes the range of crops we can grow without radically changing the soil," which is what puts forestry crops on the map, even in inner-city Detroit. "They'll do just fine here."[24]

Current plans are for a U-cut Christmas tree farm, U-pick apple orchards, and even morel mushroom fields among the orchards. The idea of hardwood forests being planted in the middle of Detroit intrigued me, as I imagined stands of oak, beech, and maple where nothing but a few sad structures were left. Perhaps Score's upbeat optimism was rubbing off on me.

A food garden has been established in a roofless shell of a brick building in Detroit. Photo © Caesandra Seawell. Used with permission.

Curious as to where a man like John Hantz would live, as well as wanting to see these islands of prosperity in Detroit, I asked Score if we could drive to Indian Village. Within minutes, we were there. Indeed, grand mansions lined a handful of streets. "This is three hundred acres with only three streets of prosperity," Score said. Indian Village is literally hemmed in on all sides by abandoned homes and vacant lots.

We had been driving through neighborhoods for well over an hour. Score had made his point. I was not getting a tour of a small pocket of blight—it was literally everywhere. "We could drive for two more hours like this," he said.

Instead, we returned to the Hantz Farms office to continue what Score called our "urban adventure" by foot. We crossed a back lane, ankle-deep in snowdrifts that showed a few rabbit and other animal tracks. Clearly nature was quickly retaking what had been a very large

Laying hens weather the winter at the same Detroit garden in an abandoned building shell. Photo by author, February 1, 2011.

warehouse or factory not all that long ago. We wiggled through a small gap in a chain-link fence and then through an opening in the metal siding of a massive abandoned warehouse. The idea is to use existing structures—such as the 45,000-square-foot building we were in—for indoor hydroponics and a market retail location. A rainwater collection system from the roof will feed into a large pond because when streets and curbing are removed, the area will be home to a reconstructed wetland area. Score has been in talks with a company that wants to cover large portions of the roof with solar panels. Another company is interested in installing a biodigester to produce methane as an alternative energy supply.

Standing inside the abandoned factory, with only a shaft of sunlight streaming through and debris strewn from various "occupants" of this space, Score summed up the goal of the farm: "We'd go from blighted residential and abandoned industrial to scenic agriculture, indoor growing systems, research on brownfield remediation, driven by alternative energy."[25]

The plans have been pored over, the outside interest and investment is poised to leap onboard, and the ideas are still on the drawing board, testing both Score's and Hantz's patience. In late 2009, Detroit elected a new mayor, businessman and former NBA point guard for twelve seasons with the Detroit Pistons Dave Bing, now in his late sixties. It has taken time for the new mayor to get a handle on just how dire the city's finances are, and to come up with a plan for urban revitalization. The plan currently on the table for revitalization is called the Detroit Works plan. It will essentially relocate the existing population, or "aggregate residents," into a few select communities, creating denser neighborhoods. In other words, move the people who are left in Detroit together. Score pointed out that this will open up even more space, making the Hantz Farms project even more appealing, perhaps essential, to city hall.

Yet, the plan to launch Hantz Farms has been in a year-long holding pattern. The city is still working out if it wants to sell land to Hantz, and if so, under what conditions. The smaller five- or twenty-acre farm site (whatever it ends up being) will be a prototype for the kind of impact a farm can have on Detroit.

"What we've said to the city is to let us show you what it looks and feels like over here," Score said, referring to the smaller test parcel. "After we've started and after most of the work has been done, you can get feedback from the people who didn't move. What's it like living next to this farm? Is the area cleaner? Is it safer? Are people saying nice things about Detroit because we've got this new farm in the city?"

"If everything that we are promising will happen, happens—that this isn't bad, it's good—let's take a look at this area. Hantz has made it clear that he'll buy as much land as the city wants to sell him. "And if another buyer wants to pay more, John won't be offended. But in essence, he'll set the floor on the price." Hantz, according to Score, has said that he'd buy ten thousand acres if the city wants to sell.

Farming, as I've come to learn, is a marginal business. It's not the way to make money, so I asked Score outright what the projected return on this project might possibly be. Depending on the models and the mix of products (meaning direct retail versus wholesale, percentage of high-value crop versus others, grant money for research on brownfield cleanup, urban-agriculture "tourism" that might result), "somewhere between minus 2 percent and a 5 percent profit," Score replied. If the project can just carry itself forward throughout the years on the initial thirty-million-dollar investment, said Score, Hantz will be happy. "It's a legacy project for him."

It's a lot for me to absorb mentally—hardwood forests and morel mushroom fields where half-crumbling houses and dump sites now stand. No wonder the politicians are moving slowly and carefully. But in the end, something has to happen to turn Detroit around. On my visit there, I heard over and over that "Detroit needs a win." That's an understatement. It needs a miracle. But Detroit has been waiting for a miracle for several decades. It's finally at the point where something radical has to happen or it will be a ghost town with a famous past and no future. Hantz might be a visionary or a profiteer; only time will tell. For now, urban agriculture is the only offer on the table. If the city gives Hantz Farms the go-ahead, the world will watch to see if Detroit becomes the most radical experiment in transitioning to a postindustrial city. And if it does, it will emerge as global center for urban agriculture.

As I flew out over the city, getting a bird's-eye view of the pock-marked urban landscape, I couldn't see any other choices. I tried to imagine orchards and forests. Moreover, I tried to imagine people.

Chapter 13

CHICAGO

The Vertical Farm

**The line of the buildings stood clear-cut
and black against the sky; here and there
out of the mass rose the great chimneys,
with the river of smoke streaming away to
the end of the world.**

—Upton Sinclair, *The Jungle* (1906)

Dickson Despommier's *The Vertical Farm: Feeding the World in the 21st Century* arrived in late 2010 to as much promotion and anticipation as a book gets these days.[1] Well before the book's publication, Despommier appeared as a guest on the *Colbert Report*, the culturally influential satirical news program on US specialty channel Comedy Central. Musician and activist Sting blurbed the book's cover. Majora Carter, a MacArthur "genius" fellow, contributed the foreword. And the *Economist* appointed Despommier "the father of vertical farming" in its magazine pages. Articles about vertical farming were seemingly everywhere at once. According to the media, the year 2010 was the year of the vertical farm—essentially a skyscraper layered with pigs, fish, arugula, tomatoes, and lettuce. There was just one problem. No one had yet built one.

Sure, there were a number of architectural renderings on paper just waiting for a visionary developer or a wealthy billionaire looking for a legacy project. Despommier's book features images of the thirty-story verdant spiraling staircase that American architect Blake Kurasek envisioned as his 2008 graduate thesis project at the University of Illinois at Urbana-Champaign.[2] It also includes the drawing for the Dragonfly vertical farm concept, an elaborate 132-floor wing-shaped "metabolic farm for urban agriculture" designed for the New York City skyline by Belgian architect Vincent Callebaut.[3] These visions were (and still are) undeniably intellectually interesting and aesthetically impressive, as are those of Despommier and fellow professor Eric Ellingsen's own glass pyramidal farm.[4] Ellingsen's work was designed with Abu Dhabi in mind, as it is likely the only city with the money to build such structures for food production. These vertical farms, however, would likely come with a $100 million price tag or more—perhaps just one of the reasons they remain more science fiction than food-growing reality.

A few years ago, not many outside academia had heard the term *vertical farm*, but the concept has been around since the Hanging Gardens of Babylon, with its mythical living walls of cascading greenery. With traditional farming being so land-, water-, labor-, and fuel-intensive, it was a logical leap to transform the two-dimensional nature of farming by shrinking its footprint radically and adding a third dimension: height. A farm built as a high-rise, with different crops or livestock layered on every floor, could conceivably allow large-scale food production right into the middle of any space-starved urban setting.

The vertical farming school of thought has led to some provocative designs. MVRDV, a Dutch design firm, proposed Pig City in 2001, an open-air forty-story farm that would house fifteen million pigs and produce enough organic pork for half a million people and endless amounts of manure for biogas.[5] It earned the vertical farm an early nickname of "sty-scraper." Other open-air vertical-farming concepts emerged soon after on architects' drawing boards in Toronto, Vancouver, Paris, and Chicago, but none were actually built.

The most recent wave of vertical-farming ideas is especially focused on "closed-loop systems." (Think a traditional mixed farm, sliced into layers, stacked vertically, and hermetically sealed under glass.) Livestock waste is intensively recycled as plant fertilizer; freshwater fish grow in tanks and produce nutrient-rich water for salad crops; water loss due to evaporation is minimized; and the whims of Mother Nature no longer interrupt the 24/7, 365-day-a-year indoor growing system. Hungry deer, grasshoppers, and other pests wouldn't devastate crops. Climate wouldn't matter—nor would climate change, droughts, or mid-crop hailstorms.

For some, this will be the only way to feed our growing cities in scenarios of nine billion people living in the megacities of the very near future. For others, it's putting the cart before the horse. Vertical-farm designers and architects talk about aeroponics (soilless growing where roots are merely misted with nutrient-dense water), hydroponics (growing plants in nutrient-rich water but without the need for soil), and aquaponics (indoor fish farming tied in with hydroponic techniques to form a self-cleansing and self-fertilizing water-recycling loop) as if we've perfected these techniques. We've been experimenting with them on rather small scales, but large-scale farming is another matter. The technology isn't there yet. Then again, Leonardo da Vinci drew models for helicopters in the fifteenth century.

What will push the technology forward? Maybe a combination of factors that are currently upon us: Climate change, rapid urbanization, the rise in fuel costs of conventional farming and transportation, and population growth may finally stretch our current food resources to the limits.

Time will tell if these models, or versions of them, will become viable as the technology catches up to the visions of the future of urban farming. For that to happen, however, a lot of ground will have to be covered. Specifically, there will have to be a significant leap in construction and indoor growing technology, especially for the fanciful vertical-farm skyscrapers in Despommier's book to leap from page into being.

Just when I thought the vertical farm was decades away from becoming a reality and that we'd continue to imagine elaborate futuristic scenarios that seemed to completely ignore that agriculture is a marginal

Chicago is home to many well-established community gardens, including the Howard Area Community Garden, which was established in 1982. Photo © Cristina Silva. Used with permission.

business, I learned of Chicago industrial developer John Edel and the new urban reuse project he's calling The Plant. It lacked the ego-driven designs of the other vertical farms that were languishing on paper, and its modesty and practicality made the idea of an indoor multistory farm seem feasible. It was enough to make me want to take a look for myself. After all, if Edel could accomplish even a modest version of a vertical farm, it would be urban-agriculture history in the making. I made plans to visit Chicago to see The Plant in its early stages of becoming the world's first, albeit four-story, vertical farm.[6]

THE PLANT, CHICAGO

As Blake Davis took off his dust mask and slapped puffs of concrete off his hands, he joked, "Clearly, as you can see, I'm a college professor."[7] Davis, a burly Chicagoan with a crew cut and a constant grin, teaches urban agriculture at the Illinois Institute of Technology. The day I met him, however, he was putting some skills to use from his preprofessorial days. His worn Carhartt® work jacket and overalls were covered in fine concrete dust from jackhammering concrete floors rotten with moisture. By afternoon, he'd be wielding a plasma torch—like a welding torch, but it cuts through stainless steel, slicing panels of it out of meat smokers for food-safe countertops and other novel reuses. Chicago had "literally, millions of square feet" of vacant, often abandoned, industrial space right in the city," Davis said. "It costs too much to tear down."

Davis was just one of several members of Edel's team of highly skilled, sustainability-minded volunteers determined to strip the former 1925-built, 93,500-square-foot (8,700-square-meter) meatpacking plant back to its outer red-brick shell and put as much of the recycled materials back into use to create a working model for a vertical farm.

While other entrepreneurs might be tight-lipped about their proto-type projects—vertical farms are the current holy grail of urban agriculture, and there will likely be significant amounts of money for those who can deliver workable models—Edel instead cleared a few hours to show me around his "fixer-upper." He let me roam at will to chat with people like Davis, Alex Poltorak (another volunteer with engineering credentials), and Audrey Thibault (an industrial designer who, as her jobs kept leaving for China, figured that she "just wanted to be part of something awesome" like The Plant).[8]

It's an experiment in motion with two rather ambitious purposes. If Edel and his team can figure out the right models and mix of elements that actually work synergistically, they will have built a viable physical and economic model for a vertical farm. Edel also intends that The Plant will serve as an open-source laboratory and catalyst for industrial reuse in a city that has no shortage of ready-built shells just waiting for a reason to remain standing.

Chicago's Stockyards

In 1878, Gustavus Swift built the first refrigerated rail car, which quickly allowed the meatpacking industry to concentrate in Chicago, scale up to incredible efficiencies, and go on to dominate the national market. By the turn of the 1900s, the Union Stockyards covered 435 acres (176 hectares) and became known as "the hog butcher to the world." If that was a slight exaggeration, it was at least the butcher that fed America. Eighty-two percent of the meat consumed in the United States at the time came from the Union Stockyards. It achieved huge efficiencies of scale that had never been attempted in livestock agriculture before. Historic photos show aerial views of the forty acres (sixteen hectares) of cattle and hog pens; what would now be referred to as a Concentrated Animal Feedlot Operation (CAFO).

The industrialized meat trade came with significant hidden costs then as it does now. The poverty, squalor, and brutal working (and living) conditions experienced by workers in the meatpacking industry were immortalized in Upton Sinclair's 1906 novel *The Jungle*. Waves of cheap, nonunionized immigrant and "underclass" labor allowed for the innovation of assembly-line slaughtering, butchering, and processing of the carcasses.[9]

The Back of the Yards neighborhood came to life as a bedroom community, if you will, for the waves of immigrants who cut and packed meat, and for the various businesses—tanneries, soap manufacturers, and instrument-string makers, for example—that surrounded the meatpacking industry on the south and west boundary of the Union Stockyards. By the 1950s, however, meatpacking was headed west, closer to the herds and where land was cheaper. The stockyards officially closed in

1971, and the only relic from that era is a giant limestone entrance arch. Back of the Yards transitioned somewhat into an industrial park. But over the years, the massive infrastructure had a dwindling reason to exist. And when industry leaves, as it did in this part of Chicago, infrastructure is left to crumble and decay. The scale of the surplus in Chicago has generally led to blight.

Much of what I saw as I left Chicago's vibrant skyscrapers and downtown core known as the Loop and made my way to the city's historic stockyards and Back of the Yards' district was heading in the direction of decay and blight. There were too many gaps in the residential streets where houses should otherwise be standing together. There were too many rusted padlocks on gates and chain-link fences encircling trucking depots, warehouses, and factories of indeterminate purposes. The businesses that remained were the signposts of a neighborhood in decline: fast-food joints, liquor stores, and convenience stores with bars on the windows.

The red-brick Peer Foods building, built in 1925 and added to over the years, was a holdout; the family-owned specialty smoked- and cured-meat company had stayed in business in the Back of the Yards until 2007.

At the time, Edel was in negotiations with the city to buy a six-hundred-thousand-square-foot World War One armory turned vacant Chicago Board of Education building. Faced with a $12 million demolition price tag, the city seemed prepared to sell it for $1.[10]

Edel already had a bit of local reputation for industrial building rehab. He had left a lucrative broadcast television design job that involved too much computer-assisted drawing and modeling to instead scratch an itch for preserving historic buildings by finding low-cost creative uses for them and reusing the materials that were simply lying around inside most of them.

In 2002, he bought a 1910 paint factory that had been officially unoccupied since the 1960s and had since become a derelict, bike-gang-ridden building with shot-out windows. (The building, in Edel's words, had been colonized by "lots of tough guys" with names like Googs, Mack, Santa Claus, the Boob, and Cowboy. There were "lots of guns, lots of knives," involved in the "informal economy" that had taken over the building.) Edel completely reformed the 24,000-square-foot (2,230-square-meter) building, putting his industrial design training, a tremendous amount of personal and volunteer sweat equity, and innate scavenger mentality into play. Useful industrial machines, like a giant, old air compressor that was left behind, were put back into service to run the air chisels used to poke holes in brick walls and the jackhammer used to remove unwanted concrete. Scrap sheet metal was refashioned to create such items as a new entrance awning, and former machine-tool parts and pipes found lying around became an art-school-esque stairway banister. Edel planted a living green roof with thousands of heat- and drought-tolerant sedum (a succulent plant that needs little irrigation) to mitigate storm water runoff and installed cisterns to catch rainwater for reuse in the building. (Seen from above, or on Google Earth®, the thousands of sedum create a red-and-green pattern of Edel's daughter's smiling face.)

Edel did it all on a shoestring budget, and 95 percent of the existing derelict structure was repurposed. The building is now home to Bubbly Dynamics, though its official name is the Chicago Sustainable Manufacturing Center.[11] Bubbly Dynamics now runs at 100 percent occupancy and is a magnet for the niche boutique manufacturing and sustainable technologies entrepreneurs in Chicago. It is home to thirty-five permanent salaried jobs, which include a co-op of five custom-bicycle-frame builders, a fabric-print-screening outfit, and a tutoring program for at-risk children. It's full and extremely efficient, and it turns a profit for Edel, the landlord. It was all the proof he needed to confirm his gut feeling that no building is so derelict that it can't be saved and made profitable.

After the success of Bubbly Dynamics, Edel's next idea was to turn another hopeless case of a building into a zero-waste organic food-producing building in Chicago. He thought he'd found it with the Board of Education building. Edel wanted to create a net-zero building that combined some select food-manufacturing processes with the growing of food.

"Everybody in city government, except one alderman, was in support of it. Instead, he wanted to tear down the 'orange-rated' historical building we were trying to acquire and have a Walmart. That was *his* dream," Edel recounted.[12] (Orange-rated is a Chicago urban-planning term that means that the building was one step below landmark protection status.)

One alderman's Walmart dream was enough to stall the process for two years, but during that time, Edel continued to plan an ambitious new life for the 600,000-square-foot (55,741-square-meter) space, using his team and networks of like-minded, hands-on experts who had gravitated to Edel and Bubbly Dynamics. That's when and how Davis fell into Edel's orbit. Davis was looking for urban-agriculture projects for his students, and Edel's business models included lots of volunteer hours and "open-source expertise." While Edel worked on acquiring space, Davis and his students began working on a symbiotic aquaponics/hydroponics system integrating fish production with a plant-growing system in the basement of Bubbly Dynamics.

Though the one-dollar price tag of the Board of Education building was attractive, the negotiations with the city were dragging on. Edel decided that ultimately it wasn't worth the wait, given all the existing inventory of available buildings in Chicago. He found a former meat slaughtering, smoking, and processing plant that was in relatively good shape. It had been built in 1925 but over the years had been upgraded and expanded. And it had sat empty for only four years, so there hadn't been time for too much to deteriorate. Most importantly, it was built for food production, which would save Edel an enormous amount of time and money because it was already up to code for many food-related commercial purposes.

Edel closed on the old Peer Foods Building on July 1, 2010, for $5.50 per square foot. What sounds like a real estate bargain, however,

Peer Foods building, Back of the Yards, Chicago.
Photo © Rachel Swenie. Used with permission.

amounted to a $525,000 purchase that would test even Edel's resource-fulness. But Edel seems just as capable of attracting paying tenants as he is overqualified volunteers. There's already a list of entrepreneurs who have signed up for space at The Plant, which will move businesses in as its space is completed.

Touring The Plant

I wasn't prepared for how shockingly cold (and dark) it would be inside The Plant on the early January day I had arranged to visit.[13] It certainly wasn't the natural-light-flooded ethereal skyscraper that the academic vertical-farming camp was known for; it wasn't even the conventional greenhouse structure one associates with a covered growing space. There were high ceilings, which on that particular day actually seemed to trap the chill, making it a few degrees cooler on the inside than it was outdoors.

I had somewhat naively assumed that Edel would have to "work around" the lack of natural light, that it was a problem to be solved. Instead, Edel explained that the thick brick walls and lack of windows was

a major benefit of The Plant. What currently functioned as windows—
antique glass block—would, however, have to be replaced. ("Glass block
neither lets light in, nor does it keep heat in or out," said Edel. As windows,
they were useless.) One of the few outside purchases that the building
would get was some new windows with high-efficiency glass.

However, high-efficiency glass is very limiting as well, explained
Edel, holding up a sample of a high-efficiency window product he had
been considering. "See how dark the glass is?" It was a smoky-gray color.
High-efficiency glass, by its very nature, blocks those parts of the light
spectrum that plants need for growing. And clear glass, which lets more
of the light spectrum pass through, allows too much heat transfer. Edel
then explained the problem of light units in northern latitudes during
the Chicago winter. "In the upper Midwest on a day like today," he
snorted, "you'll get no usable light. In an ideal [summer] day, you might
get light penetration of about fifteen feet.

"That means you'll be growing under artificial lights anyway. And
the last thing you want is huge amounts of glass for that heat energy to
escape through." Any gains made by electrical savings on using natural
light would be negated or completely irrelevant compared to the heating
costs escaping out through glass. Besides, a well-insulated brick building
such as The Plant will be very effective at trapping heat inside in the
winter (the heat from the lights can go a long way toward heating a
building if it's well enough insulated, Edel believes) and keeping it cooler
in the summer. Heat, as I would learn that day, is as valuable an asset in
an ultra-efficient vertical farm in a cold climate as anything else.

But the great advantage, Edel explained, to the cavernous nature of
the building is that "you can control the time of day." This gives Edel the
ability to "grow at night" when electricity costs are a fraction of what
they are during the day when the demand is high. And plants need a
period of darkness just like they need a period of light, so you can create
night during the day, when energy costs are high. Edel figures he can cut
the energy expenses in half by growing during nonpeak hours.

The other advantage, continued Edel, is that "you can create different time zones in various parts of the indoor system. You can flatten your nominal load so that you don't have demand spikes." Electrical utility companies like to charge you at the rate when you are at your peak daily energy consumption rate. By "moving the time of day around" between a few growing zones, again, you can achieve a "flatter," more consistent pattern of consumption and therefore save on utilities. Flattening the demand for electrical consumption will play a huge role in regulating the metabolism of the building as the building starts to produce its own electrical power and heating when the anaerobic digester is built and takes over the energy needs of the tenants and the food-growing spaces.

The one concession Edel has made to a tiny bit of inefficiency will be the "growing lobby." Large windows along the front of the building will let in lots of natural light. "We'll have things like hops and lavender, and probably the finishing tanks for the tilapia where the water is really clean and the fish look pretty."

Heat and light were not the only valuable commodities in the building's equation; oxygen and carbon dioxide also needed to be considered. Nathan Wyse, a fresh-faced twenty-something came by The Plant that day to talk to Edel. Wyse was a potential tenant who was looking to take his Thrive label of kombucha—a fermented medicinal tea hitting the lucrative mainstream specialty-beverage market these days—to the next business level.

The yeasts used to ferment the sugars in kombucha require oxygen and produce excess carbon dioxide in the fermentation process. Growing plants, handily, love carbon dioxide. According to Edel, plants "do quite well on six times the normal atmospheric carbon dioxide." Wyse asked if Edel could think about how these two gases could be exchanged efficiently between the brewing space and the growing spaces at The Plant.[14] If they could be exchanged, Wyse's aeration of his batches of kombucha would be greatly enhanced. Edel suggested that they could likely pipe excess carbon dioxide into growing areas, while drawing oxygen out (one being a heavier gas than the other) to recirculate it between the kombucha fermentation beds and the growing beds.

"So you've already thought about this?" Wyse asked.

"I just did," said Edel, matter-of-factly.

"OK, well, I'd gladly exchange carbon dioxide for oxygen for better fermentation."

I felt like I'd stepped into the future, where resources like oxygen and carbon dioxide are valued on an open-market trading system. Clearly, a closed-loop system, such as a vertical farm, as Edel conceived it, was so much more than providing artificial light to a few plants and recycling fish waste as plant fertilizer. It was about striking a delicate balance in the building to create a zero-waste ecosystem where "the only thing that will go out is food."

As we climbed the stairs to the second floor, the unmistakable greasy aroma of bacon wrapped itself around me. "The smokers were in use twenty-four-seven right up until the day Peer Foods moved out," confirmed Edel.

Some of the smokers were new: huge stainless steel tanks with what looked like ships' portholes at about five feet high. The stainless steel was valuable, and Edel and crew had already started to hack it into panels for food-grade countertops and tables. Other panels would become the new bathroom stalls.

There were also older cavernous smokers that smelled like they had been used continuously for a century, which was likely not far off. Smoke stains had left huge black licks up the beautiful 1920s glazed-tile walls. I remarked that it was a shame to think that buildings like this were decaying and being torn down due to a lack of knowledge of how to resuscitate existing construction. And yet, city aldermen had dreams of demolition and replacement with Walmarts.

"Building a new building is a really inefficient thing to do!" Edel fired back. "Plants don't care about columns, or taking a freight elevator to get out to a market. Really an existing structure is the best possible situation."[15]

The stainless steel smoke tanks were in the area designated to be the bakery, one of the food-based business incubator areas. Start-ups will be able to rent the space by the hour and still be in a completely 2,000-square-foot (185-square-meter) food-grade shared commercial kitchen, a major economic hurdle for most people getting into the food-production business, given the overhead on commercial space. Tenants can also rent garden plots on the rooftop garden and source other items, like mushrooms, that will be grown in other parts of the building. "There'll be a wood-fired oven in here," enthused Edel. The heat from the bakery will be important to heat the other parts of the building. Because of the original function of the building as a food facility, the floors undulate every few feet where floor drains exist. "How expensive would that have been to put in?"

"All of these rooms were great forests of electrical wires, pipes, and everything else. There was meat-cutting equipment everywhere. We are keeping bits of it and reusing almost everything. The oldest wiring is only fifteen years old, fortunately." There was even a beauty to the age-blackened iron rails formerly used to move the carcasses along from one worker to the next. Edel was planning to keep them suspended from the timber supports as a historical memento of the building's past.

"This is the one mess I'm going to keep because it's so out of control," he laughed, pointing toward one particularly absurd tangle of meters, pipes, wiring, gauges, and switches. Edel quipped that this is where his art school education will come into play. A floor-to-ceiling glass wall will be installed and dramatic lighting will be focused on the "industrial found art"—a ready-made point of interest that will be a central art piece on the third floor, visible from the conference room and the incubator office space that will be rented out to small businesses that will use The Plant's commercial baking, brewing, and food-preparation facilities.

The New Chicago Brewing Company has signed on to be a major keystone tenant, and there will even be a homebrew co-op that operates out of The Plant. Not only will brewing produce a lot of heat; it will supply vast amounts of spent brewing mash to compost for the gardens and greenhouse or for the biodigester.

We descended into a dark, cavernous basement for the grand finale. We cautiously picked our way around scrap metal, spools of wiring, and over curbs that were scheduled to be sledge-hammered like we were climbing through the innards of a submarine. Edel pointed out rooms that would soon be filled with mushroom beds. He had secured a former military fighter jet engine that would be put into use for electrical generation once the biodigester was built.

Edel yanked on a solid steel door, and we passed from the submarine scenario into a laboratory-white immense room bathed in a fuchsia light on one side, with gurgling vats of tilapia-filled water on the other. The Plant Vertical Farm wasn't just demolition and future scenarios; there was actual food growing in test systems in this basement room.

"This is Growing System Number One," said Edel as we walked toward the four square plastic 275-gallon (1,000-liter) tubs that were the fish tanks. This was the project that Davis's students were working on, tweaking and perfecting, so that it could be implemented on a larger scale when The Plant ramped up its food ecosystem.

Slivers of fingerling tilapia flashed around the tank, and as soon as they saw us looming over, they made for the surface. "They would eat twenty-four hours a day," said Edel, as the fish poked at the water's surface. There were two more tanks attached to this chain of plastic vats and white plastic PVC pipes, and the nearby pump was noisily forcing water around through the tanks. Sixty market-weight tilapia swirled in the final tank. "You want to control how much you feed them or they'll get too big, too fast. And you also have to balance the amount of food with the amount of plants you are growing." Edel explained that the fish were "on a diet" until they got more plants into the system.

The fish-plant loop in the basement of The Plant, Chicago.
Photo © Rachel Swenie. Used with permission.

The water from the fish pens flowed into another water-filled tank with run-of-the-mill hardware-store black plastic garden netting for filtering. Edel explained that the netting caused "the richer stuff" to fall to the bottom of the filter tank. When The Plant's biodigester is ready, this solid fish waste will be used to produce methane gas, which will be turned back into heat and electrical energy.

The next tank after the netting had a black plastic honeycomb-like panel—"a $400 mistake," whined Edel. The tank is simply a place to harbor the bacteria that turns the ammonia of the fish waste into the nitrites and nitrates (the nitrogen compounds) that make fantastic plant fertilizer. Instead of the special, expensive plastic comb, Edel proposed that "rocks or old chopped-up plastic bottles" would do just as good a job for a fraction of the price.

The pump then sent the water from the filters into shallow pans where foam rafts studded with tiny plant plugs floated on clear but nitrogen-rich water. Each hole in the raft contained a small plastic basket filled with coconut husk to stabilize the roots of each little seedling of arugula, red lettuce, or whatever the team wants to grow. The coconut husk fiber is nearly indestructible yet is porous enough to not restrict the rooting systems that dangle through the gaps in the baskets and into the water. As the plants take up the nitrogen, they effectively clean the water—as they do in ecosystems in nature—allowing the water to be recycled back into the fish tanks for the waste-fertilizer loop to begin again.

The plants looked very happy and healthy bathed in the fuchsia light of the state-of-the-art LED grow lights. "Plants can't see green," Edel explained, so you only need the red and blue lights. Edel, Davis, and students are testing the LED lights, as they are relative newcomers to the market; but if they work, they'll be much more efficient than other grow lights commonly used. A computer engineer is working out the open-source software and hardware that will move the lights along a variable-height track suspended above the seedlings. The lights move slowly from one end of the beds to the other "so they don't end up growing like this," explained Edel, listing sharply to one side.

I finally asked the big question that seems to be a sticking point where new ideas tend to hit the proverbial brick wall of city bylaws. "And you're allowed to do all of this?"

Overall, the city has just let Edel and company continue without too much concern. The brewing permit was a hassle, but they got it. "The only other resistance we'd had is from the zoning department that didn't like the idea of fish and aquaculture," said Edel. "Not for any *good* reason, because under the same zoning, you can crush cars, smelt iron, and slaughter cattle. But raising organic fish for some reason is bad. Go figure."

The fish were not yet a particular concern anyhow, as they were part of Davis's students' course work. They were working out the details of this aquaponics-agricultural loop as part of the student curriculum, which involved the microgreens, sprouts, and mushrooms that would soon be tested out at the Plant.

Part of this course work also included marketing plans and economic feasibility studies by students at the Illinois Institute of Technology (IIT). When I asked Davis how strong the demand was in Chicago for locally grown food, he replied that even drawing from a radius of five hundred miles around the city, there aren't enough farms for the markets and the demand that already exists. And being right in the city will be a huge advantage for restaurants willing to pay a premium for ultra-fresh product. "We're about the only people who can say, 'We'll pick this for you at nine a.m., have it to you by ten, and you can serve it for lunch.'"[16]

The other factor that favors the viability of vertical farming in the city, according to Davis, is that Chicago's public school system now sets aside 20 percent of its school lunch budget for local foods. "Even keeping in mind that they don't actually go to school in the summer when most of the food is produced, it still creates opportunity for us."

Food wholesale produce suppliers have also told Davis that they'll take everything The Plant can produce. So whether it is Chicago's sustainable and premium restaurants willing to pay top dollar for The Plant's fresh, local, organic food, or local produce wholesalers, or the Chicago Public School System (though clearly the school board wouldn't be able to out-compete the other two on price), finding markets for the food will be the easy part.

In Davis's opinion, however, Edel's plan of having manufacturing tenants subsidize the food-growing spaces was a key element to turning The Plant into reality while the other more ambitious "food-only" skyscrapers are lingering on paper at this point. "We've been to almost every other urban agriculture site within five hundred miles, and we noticed that almost all of them are being run on job-training grants from foundations. We thought that this was probably not a good way to run this. That's why I really jumped on to this project. It's technically interesting, but it has a commitment to creating a business model that can be replicated. The problem with social services and forty-story urban farms is that you train a bunch of people, but there are no businesses out there to hire them."[17]

When I remarked that it's somewhat surprising that the world's first vertical farm won't be nestled in among skyscrapers in uptown Manhattan, or in the anything-is-possible cities like Shanghai and Dubai; that it will happen on a very modest scale, on a very modest budget, in Chicago, Davis just smiled. "That's kind of the tension between New York and the Midwest. All the actual urban agriculture is happening within five hundred miles of Chicago, and all the press is about these forty-story buildings."

"When Sam Walton [founder of Walmart] started, he didn't try to build a four-hundred-thousand-square-foot superstore. He took an old Kresge's and said, 'I'm going to figure out this business model in this relatively small space. If it's successful, I'll make another one.' And at some point, you can afford to build a single-purpose building for a Walmart. I think if you get good at urban agriculture, and have a few technological breakthroughs, at some point you'll need an architect to design an eighty-story urban farm. Maybe your business model will be sound to do that. It's just a bit premature right now."

I asked if the city was therefore giving The Plant any breaks or help in any way. "They're not subsidizing it," answered Davis. "But the most important thing in Chicago is that they're letting us do it."

*   *   *

Edel's concept of industrial reuse seems like a reasonable solution to the very sticky wicket that has so far kept urban vertical farms confined to academic presentations and scrolls of architectural plans. And, as Edel put it, "You've got to sell a lot of rutabagas to pay for a $100 million building." Edel's ability to reinvigorate unwanted commercial space, make it beautiful, and, perhaps most importantly, make it productive and profitable once again, might just be a catalyst that will serve postindustrial Chicago well. And it might be vertical farming's Sputnik moment, launching a vertical-farm race, so to speak, that will leave those ego-driven skyscrapers on the drawing board for the time being.

Chapter 14

CUBA

Urban Agriculture on a National Scale

Use all of science for a more sustainable development that does not contaminate the environment. Pay the ecological debt and not the external debt. Fight hunger, not people.

—Fidel Castro, United Nations Conference
on the Environment and Development,
Rio de Janeiro, 1992

W hat *would* it be like if the borders were to close and cut off all food imports? What if that global food chain that supplies our supermarkets and restaurants broke down or those supplies simply went elsewhere? What would happen if there suddenly was no fuel to pump into the state-of-the-art combines, tractors, seeders, and sprayers? No replacement parts when machinery broke down. No chemical fertilizers, pesticides, or herbicides to spread on fields. And no one was willing to trade, even if you wanted to. In other words, what would happen if a sudden food shock hit an industrial, fossil fuel–dependent, globally interconnected food system of an entire nation? As Bill McKibben wrote in "The Cuba Diet: What Will You Be Eating When the Revolution Comes?"—a feature article that appeared in *Harper's* magazine in 2005 (one of the few mainstream media outlets that even acknowledged Cuba's multiyear food crisis)—"It's somehow useful to know that someone has already run the experiment."[1]

So what was the experiment, in Cuba's particular case?

From the late 1950s into the 1980s, communist Cuba followed the same path that capitalist nations did. It embraced the efficiencies of large-scale industrial agriculture, perhaps even more so in its centralized communist economy and state-owned farms. Highly mechanized monocrop farming (specifically sugarcane) and intensive livestock farming dominated. By the 1980s, 1.3 million tons of chemical fertilizer and $80 million worth of pesticides were being used on the state-directed Cuban mega-farms.[2] Sugar was king—I personally saw the legacy of the Cuban sugar mills, now rusting monoliths, in the narrow, flat midlands of the island from the window of our tour bus—with the help of copious amounts of chemical fertilizers, herbicides, pesticides, and tractors. Bulk-sugar shipping terminals could export up to 75,000 tons a day, and this one crop accounted for 74 percent of the total value of Cuba's exports in the mid-1980s.[3] The Castro government was financially and ideologically invested in high-yield, industrial farming. It was a triumph of Cuban communism that collective farms could be modern, efficient, and virtuous at the same time.

It was a false economy, however. Cuba had preferential agreements with select trading partners. The Soviet Union paid above world prices for Cuba's sugar in return for its continued enthusiasm for communist ideals and an ideological presence within 100 miles (160 kilometers) of the United States. Cuba got a sweetheart deal on fuel, machinery, rice, and wheat as it slid into the specialization trap with just a few exportable field crops. While the agreement worked, however, times were relatively good in Cuba.

In their own way, Cubans lived a very similar life to their North American neighbors: they went out to restaurants, concerts, and movies during the week; they took annual family vacations in the car to beaches on the island; and their television sets illuminated living rooms in the evenings.[4]

Domestically, Cubans were feverishly proud of their dairy industry. One cow named *Ubre Blanca*, or White Udder, achieved mythical status. You can still visit her marble statue in her hometown of Nuevo Gerona on the Isla de la Juventud. She was the most productive dairy cow in the world in the early 1980s, another achievement of Cuban industrial food production. Epic amounts of milk flowed from her teats. One day, she produced 29 gallons (110 liters) of milk—over four times the normal volume of a dairy cow.[5] It was enough to get her recognized by the *Guinness Book of World Records*. White Udder's daily production would often be reported in the national news and quoted by Fidel Castro in his speeches. The problem was that she was part of a livestock and dairy industry in Cuba that required annual imports of six hundred thousand tons of feed.[6] The tropics is no place to grow grain.

During those years of heavy agricultural industrialization, Cuba's population urbanized drastically. In 1956, about 56 percent of Cubans lived in the countryside; by the end of the 1980s, only 28 percent of the population was rural.[7] What was unique about Cuba's overwhelmingly urban population—liberated from the backbreaking work of labor-intensive traditional agriculture but without a capitalist, entrepreneurial system to invest their energies—was that they took advantage of free education, even at the university levels. Literacy and the ranks of educated citizens ballooned. This, as it turned out, would be a huge factor in how Cuba was able to survive the coming food shock.

As the communist countries in Europe's Eastern bloc started to fail in the late 1980s, and ultimately with the official dissolution of the Soviet Union on Christmas Day in 1991, Cuba had nowhere to sell its sugar. Simultaneously losing 85 percent of its foreign trade was devastating for a country that relied on foreign imports for two-thirds of its food supply, all of its fuel, and 80 percent of its farming equipment.[8] As any Cuban over the age of thirty will tell you, it's like someone turned off the lights, shut off the gas, and emptied the fridge. Cuba was adrift in a world not particularly sympathetic to a nation whose communist leader had thumbed his nose at its capitalist neighbors for thirty years running.

Without electricity, factories shut down. Tractors rusted in the fields where they were abandoned. Crops unable to be harvested began to rot. And tens of thousands of livestock starved to death on the poor native pastureland.[9]

Sensing an opportunity to finally strike a fatal blow to Fidel Castro and Caribbean communism, the United States strengthened its economic sanctions of the 1960s against Cuba with the Torricelli Act of 1992. This imposed a strict ban on trade with Cuba by any American company, including its foreign subsidiaries. Foreign ships were not allowed to use any American port if it had been to Cuba within the previous 180 days. And cash remittances by Americans to family and friends in Cuba were outlawed. It was thought that this would bring about the swift demise of the Castro government and Cuban communism. When that turned out not to be the case, the Helms-Burton Act was voted into law in the United States in 1996; its purpose was to punish foreign corporations for engaging in trade and commerce with Cuba.

Cuba was isolated from outside assistance. The most immediate concern was mass starvation. The daily caloric intake dropped by 30 percent, and the average Cuban lost thirty pounds in the three long years following the collapse of the Soviet Union.[10] Castro declared a state of national emergency and called it the Special Period in Time of Peace.[11] The austerity plan, which still lingers, was enacted to get Cuba through this post-oil, postindustrial, post-global reliance crisis. Food rationing was implemented, and the Cuban government focused its highly educated citizens' attention on revamping the Cuban food system.

The nation's scientists were tasked with helping farmers come up with high-yield growing systems that didn't require much more than human labor and cheap organic inputs. Oxen were reintroduced to work—and fertilize—the fields. Scientists developed biopesticides and experimented with companion planting and crop rotation. National seed-sharing programs were established, and state soil experts were trained and made available to any farmer who needed advice. To get around the lack of fuel for transport (and refrigeration, for that matter) Cubans turned bare lots in every city into *organopónicos*—urban organic farms with small retail shops attached, laying the foundation for a massive urban-agriculture component to anchor the nation's new urban food system. By the early 2000s, Havana was growing 90 percent of its fresh produce in *organopónicos* directly in or near the capital.[12]

What began as an emergency measure—urban agriculture—emerged as a critical cornerstone of Cuba's decentralized, deindustrialized food system, or what came to be called the Cuban Model in food-security circles.

Organopónicos, with their supply-and-demand pricing, direct-to-consumer farm-gate retail kiosks and farmers' markets, are now as common in Cuban cities as convenience stores or grocery stores are in cities in North America and Europe. There are literally several dozen urban farms teeming with vegetables, fruits, and "green" medicines (medicinal plants) that change with the season operating in even a medium-size Cuban city. Havana has close to two hundred for its urban population of two million residents.[13]

Perhaps the most shocking element of Cuba's food system to a North American or European visitor is the total lack of grocery stores, even little food marts. The urban nature of the food production and direct farmer-to-consumer nature of the Cuban food system has made the grocery store largely irrelevant. I saw just one supermarket in the diplomatic zone of Havana, and I went inside a food store in Old Havana where people were buying a few things like cooking oil, chocolate, and other otherwise rationed food. Those were the two examples of a grocery middleman I spotted in two visits. Cubans buy their food directly from urban farms or at farmers' markets and therefore eat a largely organically produced, seasonal diet that emphasizes fresh vegetables and fruit over meat and processed carbohydrates—though, to be honest, Cubans are extremely resourceful at getting access to meat and white rice, and with even the smallest affluence gains, waistlines are expanding.

Nevertheless, Cuba has emerged as a global leader in establishing ecologically sound, extremely productive, locally managed food systems driven by nutritional needs, not by profits for multinationals. And as it turns out, it's a much more sustainable and secure way to feed people. In North America, we spend between ten to twelve units of nonrenewable energy for every one unit of food energy on our plate. In Cuba, this ratio is basically reversed.[14] In 2006, Cuba was the only country in the world to achieve the targets of the World Wildlife Fund's Living Planet Report for sustainable living and development.[15] (2006 was also the year that the United States and Canada were among the ten "blacklisted" countries for their wasteful ways with energy and resources.)

Whereas an urban farm that rises from the pavement or a razed building site with dozens of types of picture-perfect varieties of seasonal, organic fruits and vegetables makes news in Canada, the United States, or the United Kingdom, in Cuba it's just the *organopónico* down the street. As a result of Cuba's unique food system, agricultural tourism has become a small but nonetheless important type of business.

It's the reason that I made two trips to Cuba. I went to Cuba in 2007 on an island-wide tour of Cuba's food and agriculture system with Canadian agricultural expert Wendy Holm. Like other people interested in urban agriculture, alternative-food systems, and food-security models, I was curious about a country that has been more or less off-limits for so long and wanted to see firsthand what an actual back-to-basics food system looks like.

I returned to Cuba again in 2010 to attend the VIII Meeting on Organic and Sustainable Agriculture international conference in Havana.[16] On both trips, though, the hardships and sacrifices of a deindustrialized food system were shockingly evident to me. Neither trip could be considered culinary tourism by any stretch of the imagination; instead, it was a frank look at what it takes to achieve a sustainable, local, clean, and equitable food system. The Cuban model attains ideals with which we are just beginning to wrestle back home: the ecological footprint of the food we consume and a transparent and secure food supply, one driven by nutritional needs, not by corporate profits. It's not easy and it's not cheap, two fundamental values that the industrialized world clings to. Nor was Cuba's transition to this model by choice. Anyone you ask who lived through this radical experiment in revamping Cuba's food system will disabuse you of any romantic notions of this period. Meat and refined carbohydrates (which Cubans dearly love) are rationed, and the selection of vegetables and fruit depends on the season. Food scarcity, twenty-plus years in, remains a daily reality for many Cubans, and between food and clothing expenses, there's little left over for even everyday luxuries, as I discovered over and over again while touring dozens of Cuba's ubiquitous urban organic farms.

THE LITTLE RADISH (EL RABANITO) *ORGANOPÓNICO*, IN THE CITY OF CIEGO DE ÁVILA, CUBA, 2007

Jorge Carmenate, a stocky, sun-weathered man in his mid-forties, welcomed our small group of curious Canadians and Americans to the *organopónico* El Rabanito, the Little Radish.[17] As he began to talk, he instinctively edged his stocky frame under the canopy of a nearby neem tree, a large tropical tree grown for the pesticidal compounds in its leaves. Our ten-person, pink-cheeked group, clutching cameras and notebooks, followed suit. Even though it was February, the mid-morning heat in central Cuba was searing.

Speaking in Spanish, with our tireless interpreter translating verbatim during pauses, Carmenate welcomed us to this seven-and-a-half-acre (three-hectare) market garden in the middle of Ciego de Ávila, a town in central Cuba that rarely sees any of the two million tourists who visit the island annually. Central Cuba's abandoned, dilapidated, and crumbling sugarcane factories on the outskirts of its towns and the Soviet-style cinderblock architecture are not exactly postcard moments. But the Little Radish is one of Cuba's top-producing *organopónicos*, and Carmenate had a well-rehearsed presentation for the growing number of foreign visitors, like us, who were interested in Cuba's urban agriculture and unique food system. He knew exactly when to pause to let our translator chime in.

The Little Radish farm was near enough to the municipal baseball diamond that Carmenate often paused to patiently wait out the roar of the fans cheering the Saturday morning game. He smiled and admitted that as soon as he was finished showing us around his garden, he would be off to cheer on his hometown team. But for the time being, he was generous with his time and clearly proud to be showing off what he and his fellow farmers had created on a flat former expanse of asphalt.

"The public decides what we plant," explained Carmenate, motioning toward raised beds dripping with fifty different vegetable crops that North American chefs would be falling all over themselves to get.[18] The ubiquitous Soviet-style apartment blocks were the backdrop to picture-perfect Chinese cabbages, tomatoes, lettuces, cucumbers, and culinary herbs, while medicinal flowers grew in raised beds that ran in neat, even strips. Ingeniously recycled concrete curbs made lovely, straight, ankle-high edges for shallow beds. Curbs were stacked two high for deeper beds, and cinderblocks were stood on end for even higher rows where needed. An open irrigation ditch ran around the perimeter, and black irrigation tubing ran to each bed with a central sprinkler. In this case, water was pumped to a cistern that would be used to create water pressure.

This is a fairly typical farmer-owned and -operated cooperative. The thirteen farmers must produce forty-four pounds (twenty kilograms) of produce to supply local schools, hospitals, seniors' homes, and daycare facilities as their "social contribution."[19] Any production above that goal is sold, free market–style, at the farm's small kiosk at the entrance. The day's selections and prices are written in chalk on a board at the little hut that serves as the point of sale. Competition keeps quality high and prices low. There are thirty-one *organopónicos* in the city of only 136,000 residents. That's one urban market garden for every 4,387 people.[20]

"Farming is hard work; you need to live in the garden under the sun all day," Carmenate continued, glancing at his thick, rough hands. "But the pay is very good and it's a very satisfying job." José, another farmer-member standing next to Carmenate, nodded in agreement and then went back to working in a row of cucumbers. The year before, for example, the Little Radish sold 227,000 pesos of produce, leaving the co-op a profit of 97,000 pcsos.[21] Half of the profit, as always, was reinvested in infrastructure; the other half was split among the thirteen workers. The land is basically rent-free, courtesy of the state; Carmenate said that they pay one peso of rent per year; that is, twenty-four cents, to the government.

One of the co-op farmer-members harvesting cucumbers at the Little Radish, Ciego de Ávila, Cuba. Photo by author, February 4, 2007.

Farmers in Cuba are at the top tier of state salaries, above doctors and lawyers, though they are far from affluent by North American standards. The state provides incredible amounts of resources to farmers to help them farm most efficiently at the lowest cost but with the most productive result. The only official occupation that seems to pay more than farming in Cuba is a job in the lucrative tourism industry.

After the economic breakdown of the Little Radish's operations, Carmenate slid into an animated technical discourse on the intricate balance of intensive crop rotations. One crop was always producing, with two others at various growing stages ready to peak as soon as the producing crop was spent. Sunflowers stood at the end of each bed to attract pests away from the other plants, and marigolds were interspersed to control unwanted bugs for "companion planting" techniques. The neem tree was not only valuable as shade in the yard but its leaves were a valuable biopesticide when soaked, macerated, and then sprayed wherever needed. A compost bin of California red wigglers were busy

The retail store at the Little Radish, Ciego de Ávila, Cuba.
Photo by author, February 4, 2007.

eating through decaying plant matter and turning it into nutrient-rich loamy soil. Cuba's legacy of industrial plantation agriculture and deforestation left the island with considerable desertification and soil erosion that continues to this day. In the cities, soil must be created almost from scratch, and the California red worms are the cornerstone of producing nutrient-rich soil to constantly replace the fertility that is lost to intensive growing and the nature of weather in the tropics (heavy rains and

Organopónico La Patria (The Homeland), Santa Clara,
Villa Clara province, Cuba. Photo by author, May 7, 2010.

hurricanes leech nutrients out of the soil and contribute to soil erosion.) Agroecology, the erudite term I heard over and over again from urban farmers, refers to a complex but effective interplay of crop rotation, intercropping, composting, vermiculture (composting with the help of worms), biopesticides, and soil management. Everything must contribute to the overall health of the garden. In that sense, the garden dictates what the farmers plant just as much as the customers do.

Produce price lists at the retail store at *organopónico* La Patria, Santa Clara, Villa Clara province, Cuba. Photo by author, May 7, 2010.

At the end of our tour—the baseball game sounding frenzied in the background—Carmenate and the other farmers gratefully accepted our seemingly odd tokens of appreciation for their time: bars of soap, disposable razors, pencils, pads of paper, deodorant, shampoo, and small household trinkets that will be shared among their families. It was a reminder that the Cuban urban-agriculture model was still a product of necessity, not an academic endeavor or a counterculture rebellion. It made me wonder how far the will to change back home would go in the presence of the same necessity and the token political support for anything but a cheap, abundant, and relatively accessible food system.

Vivero Alamar, Havana (2010)

While the Little Radish is well off the tourist path, Vivero Alamar, in the Alamar district in the eastern part of Havana city (Havana is the name of both a province and its capital city) has become a bit of a tourist attraction. It's Havana's largest and most successful *organopónico*, and now curious foreigners are commonly bused in for garden tours.

Alamar is a Soviet-era planned community in eastern Havana. The land where the *organopónico* now operates was supposed to hold a sports complex and a hospital, but it became an *organopónico* for the community in 1997. Vivero Alamar (Alamar Plant Nursery) is now a 27-acre (11-hectare) urban farm, with 143 farm workers as cooperative owners.[22]

On the day that I visited, it looked as if the entire farm had been spiffed up and staged for our group's arrival, right down to the team of oxen dragging a plow up and down rows that presumably were being readied to plant. Another team of oxen brought home a harvest of what looked like king grass, a very tall tropical grass used as livestock feed in Cuba.

Vivero Alamar specializes in vegetables and salad crops, so we saw perfect, neat rows growing in the adobe-red dirt. There wasn't a weed in sight. The farm also specializes in what are known in Cuba as medicinal

Oxen at work at Vivero Alamar (Alamar Plant Nursery), Havana, Cuba. Photo by author, May 13, 2010.

plants but what we tend to think of as just culinary herbs like mint, basil, cilantro, oregano, and the like. The farmers even grow certain plants for religious ceremonies. And the incredible, perfect produce is sold to the community at the very busy farm-gate kiosks. Vivero Alamar had enough money even to invest in black crop screens to protect most of the rows where the more delicate crops, like lettuce, grow from too much sun and heat.

Vivero Alamar, like the Little Radish, is a cooperative, which means the work is shared and so are the profits. Farmers work their way up the pay scale at Alamar, the farm's assistant manager explained to our group. You start at the bottom with some of the harder jobs, but if you stick it out, and the other farmers like you (and vote to keep you around), then you move up from a level-one farmer, all the way up to a level-five farmer. At the top levels, a farmer at Vivero Alamar will earn about $500 a year, almost twice what a white-collar job in Cuba pays, and certainly more than the $10 per month minimum-wage job pays out.

Rations

Food rationing is still the bitter reality of food scarcity in Cuba today, though there is talk that food rations may soon end if supplies can cope with an open-market system.

When I asked Cubans what items they get as part of their monthly food rations, they rattled them off so quickly that I was forced to ask several people to cobble together a reasonable list. Every Cuban can buy—at a very low price, even for them—a half-pound of cooking oil, seven pounds of rice, eight ounces of coffee, one pound of chicken, ten eggs, four pounds of sugar, twenty ounces of dried beans, one package of dried pasta, and eleven ounces of fish (which generally was not available for some reason, so chicken was substituted). Children under thirteen years of age are eligible for one pound of ground beef; children under seven get 2.2 pounds (1 kilogram) of powdered milk every ten days; children between the ages of seven and thirteen receive soy yogurt rations rather than powdered milk.[23]

Other household necessities are allotted by family: each family can buy one bar of soap every two months and one bottle of dish soap every four months, and a family of two adults and two children can buy one tube of toothpaste each month. Cubans with medical conditions have greater ration allowances. Cubans who are HIV-positive have free food rations and medications, for instance.

Cubans will tell you that these rations last for just a small portion of the month. Most food is bought at much greater expense at farmers' markets or at urban organic farms.

On my first trip to Cuba in 2007 with Canadian agricultural expert Wendy Holm, she had arranged a visit to a ration store as part of our itinerary as we passed through the central Cuban city of Camagüey.[24] Inside the store, a good-looking man in a new polo shirt and pressed khaki pants sat at a desk tucked inside a street-level entrance in a ram-shackle building. Behind him were a number of hand-drawn signs: "Cig-arettes will not be sold to minors," the noticed read. "Abusive language will not be tolerated." There were a few dozen dinner rolls in the front display case, but everything else was kept out of sight. Only sample items were displayed with hand-written tags identifying the price. It was about as high-tech as a lemonade stand, but in a strange way, that seemed entirely appropriate. No one was going to "impulse shop," and brands of cooking oil didn't have to dazzle to compete with other brands. The ration store was really just a neighborhood distribution hub for a few basic items, none of which were particularly of good quality or exciting. To dress up the shop might have even been a bit insulting. (Many Cubans I spoke to complained bitterly about the low quality of the rationed food).

The store was small, and I felt uneasy just looking around. It wasn't just that there was not much to look at; I knew the clerk was aware of how pathetic the selection must seem to a foreigner. We were both uncomfortable about the situation, so I asked if I could snap a picture. He shrugged acceptance. I took a quick photo. After I left, it struck me that the ration store was the closest thing I saw to a grocery store the entire visit to Cuba.

FARMERS' MARKETS

In 1994, farmers' markets were officially legalized in Cuba, in a further attempt to spur food production and lower prices for people.[25] These began as state-sanctioned farmers' markets where state farms could sell any surplus produce, but the quality of the goods was so low that soon Cuba had to authorize another type of farmers' market that operated unregulated by the state and simply by the laws of supply and demand. There are still state-run farmers' markets, just as there are state-run farms

Pepper vendor at a "free-market" farmers' market in Camagüey, Cuba. Photo by author, February 6, 2007.

in Cuba, but they are generally associated with lower-quality food items, though the farmers who work on these farms are trying to change the image that Cubans have of their products.

I visited a state-run farmers' market in 2007, and it was truly an uninspired affair. People will shop there because the prices are regulated by the state, and so prices are often lower—as is the quality. The vendors don't "hustle" their products, and the little kiosks are rundown and smack of Soviet-style severity.

The more popular, and much more common, free-market farmers' markets in Cuba are not all that different from those in North America except in a few ways. First is the eye-popping abundance and variety of tropical fruit. At the bustling Agromercado 19 y B (named as such because it's on the corner of roads 19 and B) in Havana, I bought a mamey fruit, an ultra-sweet and soft tropical tree fruit the size and shape of a large, fat mango.[26] From its rough brown exterior, I could have never guessed at the tooth-ringing sweet white pulp inside, with a flavor and consistency that I can only describe as a cross between cooked sweet potato and a very sugary avocado. But the market offered a selection from multicolored habanero peppers (not the ultra-fiery ones, since Cubans have a near-total aversion to even the mildest heat in their food) to daikon radishes, bok choy, parsley, bananas, long green beans, tomatoes, onions, limes, sweet peppers, and potatoes, among other foods.

Most North Americans and Europeans would bristle at the sight of the meat section at a Cuban farmers' market. Giant carcasses of beef and pork are splayed on open-air countertops, and usually a strong man with a big, not-the-cleanest machete hacks off cuts of meat and hands them to customers as they order. (In Cuba, you also need to bring your own plastic bag or container for whatever you are going to buy, meat included.) More than once, I saw raw chicken thighs and even whole, raw, plucked and cleaned chickens strapped down by bungee cords on the newspaper rack of a bicycle as a person left the farmers' market. It explained why meat in Cuba was almost always stewed or deep-fried.

If you are buying food at a farmers' market as a foreign visitor—and it remained unclear even to me as to whether this was "allowed"—you will pay in CUC, the Cuban convertible peso, not in national currency. In that sense, Cuban farmers' markets are not entirely on the free-market system, but knowing what Cubans make and how much they spend on food purchases, it needn't be.

Lastly, farmers' markets are noisy affairs in Cuba. Sellers call out their products to the shoppers. Some burst into song or do bird calls. But the energy and vibrancy that is on display on the morning trip to a farmers' market in Cuba is intoxicating. You just don't get that pushing a cart up the aisle at the grocery store in Canada or the United States.

**Market vendor at the "free-market" farmers' market
Agromercado 19 y B, Havana, Cuba. Photo by author, May 15, 2010.**

TITO NUÑEZ: THE REVOLUTION
WITHIN THE REVOLUTION

The best meal I experienced in Cuba was one I didn't even eat. The same day that I was able to tour the famous Vivero Alamar *organopónico* in Havana, a few of my colleagues from the VIII Meeting on Organic and Sustainable Agriculture 2010 conference went to Las Terrazas and visited the country's leading chef in the sustainable food movement eco-restaurant, Tito Nuñez.

Las Terrazas is an eco-resort village thirty-one miles (fifty-one kilometers) outside of Havana. The area was deforested by colonial coffee plantation owners (or rather by their slaves) during the brief but impactful French colonization of Cuba from 1791 to 1809. Las Terrazas was finally reforested and restored in the late 1960s. Over six million trees were reestablished there and a small eco-village was built. In 1986, Las Terrazas became a 12,355-acre (5,000-hectare) UNESCO Biosphere Preserve, and now it's a major internal and foreign tourism draw for its hiking trails, swimming, lush natural environment, and biodiversity. It is also the head-quarters from which chef Tito Nuñez is slowly creating a revolution within a revolution. Nuñez, one of Cuba's few vegans, runs a small thirty-six-seat eco-restaurant called El Romero (The Rosemary) at Hotel Moka.

Because a number from our group couldn't visit his restaurant in person, Nuñez came to Havana. He is a small-framed, monkish man with shorn hair. He also has the patience of a monk. Nuñez became a vegetarian for health and ethical reasons, and for two decades he has been patiently and slowly converting people, one meal at a time.[27]

The restaurant's motto is "so that the cows, the chickens, the lob-sters, the jutias [small Caribbean goats], the billy-goats and the fish may live." He is also a member of Slow Food International—one of the few members in Cuba. He's also deeply concerned about the health of the Cuban people, who are now suffering from the same chronic diet-related diseases that are at epidemic levels at home in North America. It was his own poor health as a child and young man that led him to vege-tarianism, and it was as lonely a path then in Cuba as it is now.

Nuñez showed slides of his eco-restaurant and the 1,076 square feet (100 square meters) of botanical and food gardens that supply his restaurant with 70 percent of the food he uses. The restaurant uses other sustainable technology such as a solar water heater to supply the dish-washer and solar cookers and wood-fired ovens to cook the food. His commitment to the environment goes beyond just onsite gardens and vegetable- and fruit-based foods. At El Romero, the staff pack the take-away items in banana leaves; the menus are made from recycled paper and tied with natural fiber; drinking straws are hollow plant stems; the wine buckets are sewn from palm leaves; and onsite beehives provide the sweetener used at the restaurant.

In addition to the Eden-like setting for his restaurant, Nuñez's technical skill was evident as he flipped through the slides of the dishes. The food presentation at his restaurant outstripped anything I'd seen in Cuba, ever. (He was trained as an industrial engineer, and his precise mathematical brain obviously informs his cooking. "What a painter does with colors," he told me, "I do with flavors.")

The food looked fresh, vibrant, packed with flavor, and completely original.

Despite Nuñez's seeming ease with this type of cooking, the vegetarians at the conference had a terrible time getting a palatable meal or even adequate calories, whether at the premium Hotel Nacional Havana or anywhere on our travels. The concept confused every restaurant chef we came across in Cuba. Rather than just put together some black beans, rice, plantains, and vegetables, they would invariably panic and produce horribly misguided vegetarian "creations," such as mounds of flavorless steamed carrots or overcooked green beans. Nuñez admitted that his crusade for healthier, more sustainable, and ethical food choices has been a tough sell to the majority of Cubans.

"This is a cultural problem," Nuñez offered by way of explanation.[28] Cubans want to eat only "meat, refined grains, and rum." But if anyone can convert committed carnivores, Nuñez and his staff seem to be in a good position to do so with a sixty-six-item menu, all of which looks like edible art.

LESSONS FROM CUBA

In Cuba, I saw some of the most impressive ecologically integrated farms in all of my travels. Both trips were dizzying experiences of how "high-tech-low-tech" devices and systems could be put into use with a little ingenuity. Osvaldo Franchi-Alfaro Roque on his farm La Joya just outside of Havana proudly showed off his ingenious irrigation timer that made use of large soda bottles, recycled hospital IV tubing, a regulator valve, and a swiveling cradle. The irrigation timer was for the seedlings in his fruit-tree nursery. Other farmers fashioned handmade hulling and husking machines from automotive and farm equipment parts. After just two weeks in Cuba, it was hard to return to the sheer waste that is North American consumerism.

I was struck by the changes that took place between my visits in 2007 and 2010, especially because of the transition of power between Fidel Castro and his younger brother, Raúl. Cuba seemed to be moving more toward socialism while distancing itself from communism. The austerity that I witnessed in 2007 looked like it was easing somewhat. By 2010, the free-market concessions and Cuba's oil-for-doctors arrangement with Venezuela were clearly making an impact. People had better and more stylish clothing. Cubans seemed generally more affluent (though that sounds a bit strange, given their relative lack of disposable income). There were more cars on the roads and fewer donkey carts. Fewer hitchhikers were on the highways. (It is illegal for a Cuban vehicle to not pick up a Cuban hitchhiker at approved highway hitchhiking areas, which are manned with official hitchhiking organizers to coordinate the flow of transportation.) I even saw a few tractors being resurrected on farms in 2010, something that was unthinkable even a few years ago.

There were a few constants, as well. The Cubans I spoke with in 2007 and in 2010 universally hoped for an end to the US embargo but feared the change that a more consumer-oriented culture would have on their quality of life. The farmers I spoke with, however, were convinced that a return to industrial agriculture would not be tolerated, either by the state or by the people of Cuba. The lessons of the early 1990s were still too painfully close. Instead, Cubans were excited about the innovations they were working toward in permaculture; that is, in designing agricultural systems that integrate with living spaces in a way that resembles bountiful, regenerative natural systems. And they were excited about integrating alternative energy such as solar and biogas into the agricultural landscape.

My travels in Cuba and the opportunity to meet several dozen urban Cuban farmers forever changed how I look at cities. I see food-growing capacity everywhere now. Cuba opened my eyes to how deeply ingrained our consumerism is. It's the tail that wags the industrialized world's food system. The Cuban Model can teach us a lot about what a more balanced, resilient, sustainable food system looks like. The question is, will enough of us watch, listen, and learn, or will we risk going hungry? We are not just faced with peak oil but with peak water, peak land, and even peak micronutrients like phosphorus. All of these are necessary to continue with industrial agriculture.

Chapter 15

CONCLUSION

Greening and Eating Our Cities

You don't have to go back to the land, you're already there.

—from personal interview with Ron Berezan, the Urban Farmer, 2009

When I started to write this book, I set out to answer a few simple questions. Why the overnight interest in urban food gardens, urban chickens, and urban beekeeping? And what else was happening in cities that were taking back control of their food supplies and systems?

Having visited as many places as I could until I simply had to stop looking and start writing, I'm now left with the task of pondering the future of food and our cities. Where will these sparks of the food revolution take us? Will our cities look like variations of one of the Cuban cities I saw in 2007, or will we build forty-story vertical farms to feed our growing urban populations? Is there enough critical mass and commitment to continue this urban food revolution, or will large cities spiral down to become extreme food deserts?

Of course, I don't know the answers to these questions. No one does. But what I have seen has given me hope that at least we are collectively heading in the right direction. There's a tremendous amount of energy being poured into greening our cities by private citizens, small and large businesses, and municipal governments, and it is starting to yield some impressive innovations and results.

The city of New York is now home to at least two major commercial rooftop farms. Brooklyn Grange Farm is a 40,000-square-foot (3,716-square-meter) organic vegetable farm on a rooftop in Queens.[1] Eagle Street Rooftop Farm, also in Queens, covers 6,000 square feet (557 square meters) of flat roof and supplies organic produce and honey to area restaurants, to the farm's CSA subscribers, and to its onsite farmers' market.[2]

Lufa Farms, a 31,000-square-foot (3,000-square-meter) greenhouse farm that began producing twenty-five varieties of vegetables in 2011 on a rooftop in Montreal, Quebec, is leading the way for year-round rooftop production.[3] The company claims that it produces enough fresh food to supply one thousand Montreal families with weekly baskets of lettuce greens, tomatoes, eggplant, culinary herbs, and even spices.[4] Also in the summer of 2011, Gotham Greens in Brooklyn, New York, announced its first harvest from its 15,000-square-foot (1,393-square-meter) rooftop greenhouse.[5] It aims to produce eighty tons of fresh produce year round.[6]

At street level, homeowners are digging up their lawns for home-scale veggie gardens. Entrepreneurial urban farmers are seeding and weeding piecemeal plots for commercially grown urban produce to be sold at a farmers' market near you. Outlaw urban chicken keepers are winning approval for their backyard flocks of laying hens. And municipal governments are catching up with the grassroots movement by starting to support more community gardens and plant public orchards in lieu of ornamental landscapes.

Schools are incorporating culinary education into their curriculum with the spread of school food gardens at all levels of education. The Los Angeles Unified School District (LAUSD)—the second-largest school district in the United States with its nine hundred schools—is making great strides toward becoming the nation's greenest and healthiest.[7] Despite budget stresses, it supports one hundred schoolyard food gardens in an effort to teach students about healthy food, growing techniques, and basic food literacy.[8] (The LAUSD also deserves recognition for being an early initiator of soda sales bans in 2003 and for bans on other junk foods in its schools in 2004.) Universities and colleges are even offering degrees and certificates in urban agriculture and urban food-security issues.

Many city halls (Colorado Springs, Portland) and government buildings (the White House) now have demonstration food gardens and keep bees (Chicago, the White House) as a show of how with-the-times they are. More importantly, they are loosening the restrictions on zoning to encourage food production in residential, industrial, *and* commercial zones. Some cities, like Toronto, Ontario, have begun to mandate green roofs on new commercial buildings.[9] Urban beekeepers are winning the battle of public approval for their "hobby." And, of course, there are a number of new television programs, books, and magazines on urban-farming techniques and memoirs.

This is certainly encouraging news, right?

For those who have been working in urban agriculture for many years (and in a few cases, many decades), they know that the excitement level is high right now, but they also have a realistic view of what it will actually take to change our habits, behaviors, and expectations of how we will live if we want to truly address sustainability in our lifestyles.

"Where do you find the time to grow your own food? Any type of sustainable living takes time," warns Ron Berezan.[10] Berezan is also known as the Urban Farmer, and since 2004 he has been a full-time urban-agriculture consultant, landscape designer, and educator. He's in perpetual motion at his many workshops on how to incorporate food into residential and city landscapes in the service of filling the skills and knowledge gap for those who want to lead more sustainable lives but who are generally one and two generations off the land, as most city dwellers are. "I'm biased," Berezan says, "but I can make a pretty strong argument that if you want to do one thing to lessen your environmental impact, grow your own food."[11]

However, Berezan is well aware that the time it takes to grow food and the physical effort involved are major barriers in our time-pressed, smooth-handed urban world. This is where permaculture and edible forests come in for Berezan. He is one of a few maverick urban-agriculture thinkers who are venturing beyond the usual urban-agriculture paradigm of raising planters where lawns once were and rethinking the model of sustainable food production. Permaculture, Berezan says, is "an ecological design methodology for determining how we can meet *all* human needs—including food—but also energy, water, shelter, and how we can do that in sustainable ways by designing these systems based on ecological design principles."[12]

At first glance, permaculture techniques tend to yield wild, unruly, seemingly unkempt gardens, but designed correctly, they produce a tremendous amount and variety of food for a fraction of the effort of a traditional garden and with greater biodiversity that incorporates flowers to attract pollinators, water sources for pollinators and small wildlife, and sunflowers that germinate where the birds drop a few seeds.

In other words, permaculture principles are designed to create or foster productive landscapes that provide food and other needs in a way that is more self-sustaining and self-regenerative. Permaculture attempts to imitate natural ecosystems. Rather than the arduous task of domesticating food plants and animals, a permaculture gardener builds systems that mimic nature's ecological balance. The designer then steps back a bit and lets nature do the work. It is sometimes explained as reaping what you do not sow. It also acknowledges that modern life and sustainable living need to meet somewhere in the middle. Permaculture is sustainable not only ecologically, but as an activity it's a *sustainable* sustainable system.

Edible forests (also called forest gardening) incorporate a permaculture concept that builds on the fact that forests and woodlands tend to produce an abundance of food such as berries, tree nuts, mushrooms, fruit, tubers, and root vegetables, for starters.

According to the United Nation's Department of Economic and

Social Affairs and the World Bank, "about 1.6 billion people depend heavily on forest resources for their livelihoods, including 60 million indigenous people who are fully dependent upon the forests and an additional 350 million who depend on them primarily for income and subsistence."[13] Therefore, it's also important to note that forest gardening, like permaculture, is not a new or Western contribution. These old principles, forgotten during industrialization and only recently rediscovered, never dropped out of practice in many other societies that sustained themselves on a much smaller ecological footprint. And edible forests are part of an idea from which hunter-gatherer food cultures have never departed.

Edible forests imitate forests and woodlands, taking food-producing levels into account: tree and bush fruits and berries, tree nuts, ground-level food and fruit plants, roots and tubers, and climbing vines. By considering the growing levels in three dimensions, rather than just two, the food yields increase tremendously. It's what vertical farming is to land-intensive surface agriculture.

The interest in edible forests and permaculture workshops has grown substantially since Berezan started offering them. Not only are these methods more resource-wise, giving more output per square foot than other growing systems, but by letting natural systems play out, there is less work for the grower. Annual food crops self-seed at the end of the season. Perennials die back and provide a layer of new organic matter to compost into the soil. Trees provide shade and protection for tender crops. Animals like chickens devour dandelions, scratch at the soil, control bugs and pests, all the while providing a mobile fertilizing service.

I start to imagine the wooded parklands in most North American cities that could be dripping with free, self-regenerative foods. Most cities, however, do not allow foraging on public lands, and some cities, like New York, are even considering banning it outright or handing out fines. Foraging in Central Park seems to have become a trendy activity, and park officials are finally saying that it has gotten out of hand.[14] I don't think we'll be sourcing much food from public edible forests in North American cities anytime soon.

Nor do I think North American cities will do away completely with the supermarket, nor will they suddenly sprout an urban farm for every five thousand citizens. But a few SPIN farmers to complement the few sturdy farming families who have managed to stay on their urban farmland is enough of a start to change people's minds about how they get their food, and from whom. And it's the beginning of moving toward a network of urban and regional farms. As long as there is sustained interest in local food, entrepreneurs will rush in to fill the demand in the market.

The fact is that there will be no singular urban-agriculture solution that will work for every community and every city. Each city will have to address its own limitations and needs: We won't go back to living in the forests any sooner than we'll be able to achieve total food security through urban farms on each corner.

If I can allow myself a few predictions at this point, it is that balance will move slightly more back to a middle ground. We won't completely give up the benefits and flavors of the international food trade and industrial food production, but we won't be at its complete mercy either.

If we learn to give space in our urban settings to food production and food producers, we'll be healthier, happier, and more connected to the physical realities of our short existence because of it. We've still barely scratched the surface of ballooning healthcare budgets that are directly influenced by our food choices. We could truly be pennywise, whereas now we are being pound-foolish.

We need to put an end to our strange fascination with trying to outwit and dominate nature, and start appreciating it. And we'll start to look at feeding ourselves and one another as a basic human right. If we can even begin to walk along the road to any one of these goals, the various experiments and initiatives in urban agriculture—from a few beehives on a rooftop in Paris to a fight over a dozen acres of farmland in a city—will have been worth every last moment of effort.

I have come to see, perhaps a bit romantically, that the urban farmer or backyard food grower has taken up the mantle of what was once the small-scale farmer: a deliberate keeper of the open-source technology that agriculture has always been, a rebel stand against the artificial idea that corporate interests can fiddle a gene or two and claim ownership of a technology that has existed for longer than we can measure.

And I began to grow more and more convinced and hopeful that nature was already one hundred steps ahead of humans, that it has so much built-in redundancy and biodiversity (specifics like bluefin tuna and cod aside) that overall, it can repair itself despite our human efforts to engineer scarcity.

"Your Garden." An example of a food garden planted by a private citizen on public streetside boulevard land, specifically for public use and enjoyment, Edmonton, Alberta, Canada. Photo by author, June 30, 2010.

Cities—or, rather, those of us who live in cities—can no longer just be consumers of food and producers of waste. We're realizing that it's time to close the loop. We need to grow some food, re-localize our diets, and compost our food waste. And by doing these things, we liberate ourselves from that *Titanic*, the sinking global industrial food system.

ACKNOWLEDGMENTS

Writing a book is a humbling marathon. I could not have done this without a large network of support and assistance from family, friends, colleagues, and (to my delight and surprise) complete strangers, who took the time to answer questions or provide information, just for the love of urban agriculture and good, locally grown food. To thank everyone who helped me get from the initial idea to the final book would be too much, but every last piece of advice, interesting fact, helpful connection, and word of encouragement to keep writing helped make this book become a reality.

One of the greatest pleasures I got while researching and writing this book was to meet so many passionate people in urban agriculture, food systems, and urban food gardening. I will always treasure the new friends I met along the way who gave their time and energy toward helping me understand why they do what they do. Each and every one, in her and his own way, helped me realize that while growing a few heirloom veggies in the front yard, tending a community orchard, or keeping a beehive on a condo rooftop may seem like an insignificant thing on its own, it is these little actions of self-reliance and community self-sufficiency that are at the forefront of the new food revolution. More importantly, the world is a richer, and tastier, place for your important work. To them, I simply say: keep growing.

There are a few people I wish to acknowledge by name: Mifi Purvis and Craille Maguire Gillies for being my sounding boards, especially during the early rumblings of this project; Adria Iwasutiak, the supreme literary networker; and my sincerest thanks to my agent, Chris Bucci at Anne McDermid & Associates, for being in love with the idea for the book from the beginning and trusting that I could see it through its long journey. And to the entire team at Prometheus Books, which shepherded this book from draft manuscript to beyond the bookshelf, my eternal gratitude for your guidance and contributions.

Above all, I need to thank my husband, Mike. This book couldn't have been completed without you shouldering more than your half of the load and allowing me the time to write. And for the daily offerings of help, which usually meant delivering plates of food and countless cups of tea to my desk, I love you.

Lastly, I got a significant lift during the final months of preparing this manuscript when I got a phone call letting me know that I was the winner of the 2011 Dave Greber Freelance Writers Award, a Canadian book and magazine prize that helps independent writers bring social justice issues to light. Thank you for getting the connection between food and social justice, and for finding my book worthy of the prize.

GLOSSARY

biofuel. An alternative fuel source made from plant-based alcohol (bioethanol) or vegetable oils (biodiesel). This is an alternative to fossil fuels, which are derived from oil and gas deposits.

colony collapse disorder (CCD). The distressing phenomenon in which worker honeybees suddenly disappear from a hive. The causes of CCD are still poorly understood. It can account for significant, and distressing, losses every year for commercial and amateur bee-keepers.

Community Supported Agriculture (CSA). A business contract between CSA-participating consumers and a farm, in which consumers pay in advance in the spring and receive a share of the harvest, usually weekly, until the end of the growing season. CSA shares are usually delivered to customers via a weekly box delivery or pickup point.

companion planting. When a biological or structural symbiosis can be used in two or more plants to increase the yield or vigor of food crops.

concentrated animal feedlot operation (CAFO). A facility holding a large number of uniform livestock in relatively close quarters for feeding, fattening, and, usually, slaughter. There is little consideration given in CAFOs to the animals' quality of life and comfort.

co-producer. The term used by Carlo Petrini, founder of the Slow Food movement, to refer to consumers. Petrini urges consumers to think of themselves as co-producers to encourage them to acknowledge that their food choices are important and that they take an active role in the food system as opposed to a passive role.

edible forest. The practice of cultivating food gardens that mimic or utilize natural woodland ecosystems as their model. This system is a lower-input alternative to food cultivation than agriculture, but it can produce surprising amounts of edible products. Also known as forest gardening.

food desert. Urban phenomenon in which, despite reasonable levels of population density, there is a scarcity or absence of grocery stores and markets that sell fresh, nutritious food. Often people living in food deserts are prevented from accessing healthful food because of lack of affordable and convenient transportation options. Food deserts can often contain fast-food outlets and convenience stores as the only option for food purchases.

food mile. The distance food travels from where it is grown or raised to where it is ultimately purchased by the consumer or end-user.

food policy. A directive that proposes guidelines and policies surrounding the production, distribution, and consumption of food for a defined group of people. Many municipal governments are adopting food policies written by grassroots food-security organizations in their cities.

food-policy council. The organizing body that writes or oversees the food policy. A food-policy council can either be an officially appointed group of people or a self-appointed citizen volunteer group.

food security. A catchall term for the level of accessibility to fresh, healthy, nutritious food for a person, family, or community. The United Nations Food and Agriculture Organization's most current definition of food security is when "all people, at all times, have physical, social and economic access to sufficient, safe and nutritious food which meets their dietary needs and food preferences for an active and healthy life."

foodshed. Describes the complex interaction of relationships between all the food that is produced, distributed, and consumed within a geographic region.

food shock. A sudden, drastic interruption to food supplies for any reason, such as a weather or natural catastrophe, war, or lack of fuel.

food sovereignty. A term describing a person's or group's ability to choose their own foods and agricultural system of production. The idea came about as a reaction to the globalization of food and the industrial food chain that leaves little cultural and individual freedom of choice of what or how food is produced.

French Intensive Agriculture. A system of food production that brought together a number of techniques such as raised beds, heavy dressings of compost, cold frames for early forcing of crops, thick stone walls surrounding gardens to create a warmer microclimate, companion planting, and densely planted crops to reduce water loss and weeding. French Intensive Agriculture allowed Paris to produce impressive amounts of fresh food in an urban setting in the first half of the nineteenth century.

Green Revolution. The term applied to the advances in crop and chemical sciences that produced much higher-yielding cereal and grain varieties in conjunction with new chemical fertilizers and pesticides. It also refers to the exporting of this new technology to at-that-time-developing countries like Mexico and India to industrialize their farming as a form of aid and foreign development. The Green Revolution took place between the 1940s and 1970s.

guerilla gardening. A form of social and political activism in which gardeners plant flowers or food on underutilized private or public land without the consent of the owner.

industrial food. Food grown and produced according to an industrialist mindset that is concerned with mechanizing and lowering cost of production by large-scale production and by concentrating manufacturing and distribution chains to achieve the lowest possible cost of unit production. Author Michael Pollan offers this definition of industrial food in *The Omnivore's Dilemma*: "Any food whose provenance is so complex or obscure that it requires an expert to help ascertain."

intercropping. When two or more crops or plants are grown simultaneously in the same area, usually to achieve a more rapid succession of harvests than if one crop was planted, allowed to mature, harvested, and then removed before the next crop was planted.

locavore. A person who seeks out locally grown or locally produced foods.

organopónico. A small-scale, organic, cooperative-run urban farm in Cuba that grows food for its surrounding community. The term is now being used for similar small-scale urban organic farms in other Latin American countries.

peak food. A term that mimics the idea of peak oil—oil being a finite resource, whereas food is generally seen as a renewable resource. *Peak food* refers to the point at which a system (global or regional) is producing at its maximum yield, after which production starts to decline. This decline would be due to soil erosion, nutrient exhaustion in the soil, water stresses, and any number of variables that would adversely affect the ability to grow food.

peak oil. The point at which the extraction of oil reaches its maximum rate and efficiency, after which the supply goes into a terminal decline and the effort (and cost) to extract it invariably rises.

peak water. The high point at which we are using renewable freshwater resources faster than they can naturally regenerate. This will lead to scarcity and higher cost as freshwater supplies decline.

permaculture. An approach to living and growing food in a permanently sustainable way that draws heavily on or even closely mimics natural ecosystems. It takes an integrated approach toward food-production systems and other human needs.

Slow Food. An Italy-based grassroots food movement that now counts over 100,000 members in 130 countries. It began in 1989 as a countermovement to the globalization of fast food but now works on issues of food sovereignty and biodiversity preservation and in service to its motto of "good, clean, fair food for all."

terroir. A French term, borrowed from viticulture and wine production, that encompasses the cumulative effect of geography, geological variations, and climate on the flavor characteristics of foods. *Terroir* is used to explain why the same crop grown in different places will taste different despite the fact that it may be the exact same crop plant—even a genetic clone, as in the case of wine grapes. It not only describes the effect of place but takes into account the time frame during which the product was made.

urban agriculture. The act of growing and, according to some definitions, distributing food in a defined urban area. The idea is to produce food closer to where the majority of consumers are, in cities. Urban agriculture tends to be small-scale agricultural enterprises, even on a household scale.

vertical farming. A concept of stacked agricultural production, in vertical layers, to shrink the footprint of land-based agriculture. Some vertical-farm designs are essentially food-producing skyscrapers (though none of these designs have been realized yet). Vertical farming tends to incorporate closed-loop agricultural systems where nutrients in food production are recycled from one food-production area (such as freshwater fish tanks) to another (such as salad greens).

RESOURCES FOR
URBAN AGRICULTURALISTS

*F*ood and the City, as you by now have discovered, is a description of the urban-agriculture revolution happening in many cities in North America and Europe. If you are inspired to start growing even a small amount of your own food (which I hope you are), I have compiled some resources to get you going.

BOOKS

City Farmer: Adventures in Urban Food Growing (Vancouver, BC: Greystone Books, 2010) by Lorraine Johnson is peppered with advice and tips on urban food growing to keeping urban chickens.

Square Foot Gardening: A New Way to Garden in Less Space with Less Work (Emmaus, PA: Rodale Books, 2005) by Mel Bartholomew is a classic book on growing food on a home scale. It's a great guide to help even the most novice gardener begin a successful garden plot.

Urban Farming: Sustainable City Living in Your Backyard, in Your Community, and in the World (Irvine, CA: Hobby Farm Press, 2011) by Thomas J. Fox is a new encyclopedic guide to urban food growing. It contains practical lists of how to build urban gardens as well as growing advice.

ONLINE RESOURCES: NORTH AMERICA

Commercial Urban Farming

SPIN farming is an international movement, thanks to an easily reproducible small-scale urban-farming model that offers low start-up costs and good profits for urban farmers: http://www.spinfarming.com.

Young Urban Farmers is a Toronto-based company that helps people grow food in cities: http://youngurbanfarmers.com/.

Community and Food Security

The **Community Food Security Coalition** (CFSC) is a North American coalition of diverse people and organizations working from the local to international levels to build community food security. The listserv community is especially responsive and helpful. Community Food Security: http://www.foodsecurity.org.

The **Detroit Black Community Food Security Network** is a community-based food-advocacy group that promotes fair-food policies, healthy eating, urban agriculture, and job training in urban food growing and food systems: http://detroitblackfoodsecurity.org/.

Growing Power, Inc., of Milwaukee, Wisconsin, is an excellent example of community-based nonprofit urban farming and community education and job training: http://www.growingpower.org/.

Community Gardening

The **American Community Gardening Association** is a binational nonprofit that supports community gardening in the United States and

Canada: http://www.communitygarden.org/. The association also has links to many municipal community gardening bodies in North America at http://communitygarden.org/connect/links.php.

Municipal horticultural societies and associations are also excellent references for urban gardening in many major cities.

Food Safety and Food Security

Award-winning author and educator Marion Nestle's excellent *Food Politics* blog is a reliable and accurate resource for food safety and food-security issues in North America: http://www.foodpolitics.com/.

La Vida Locavore is a politically charged blog headed up by Jill Richardson, author of *Recipe for America: Why Our Food System Is Broken and What We Can Do to Fix It* (New York: Ig, 2009), http://www.lavidalocavore.org/.

Politics of the Plate is author Barry Estabrook's excellent blog that covers food safety and politics: http://politicsoftheplate.com.

General Urban-Agriculture Information

City Farmer's website is the global go-to site for the latest news about interesting projects and news from the world of urban agriculture: http://www.cityfarmer.info/.

Deconstructing Dinner is an excellent Canadian radio program and Internet podcast that often focuses on leading-edge issues in North American food security, urban agriculture, and local-food systems: http://kootenaycoopradio.com/deconstructingdinner/index.html.

Grist is an online environmental newsmagazine that reports on urban agriculture and other food issues regularly: http://grist.org.

Land-Sharing Websites

Landshare is a match-making website for people who have land to lend to those who don't have land but wish to grow food, mainly in urban areas. It began as a program in the United Kingdom (http://www.landshare.net/) but has now launched in Canada: http://landsharecanada.com.

Shared Earth is the major match-making website for those looking for free urban grow spaces in the United States: http://www.shared earth.com/.

Residential Urban Food Gardening Resources

Ron Berezan, the **Urban Farmer**, has a website full of sustainable living resources, permaculture design, and educational exchanges in Cuba: http://theurbanfarmer.ca/.

School Food Gardens

Famed Bay Area restaurateur Alice Waters has put considerable time and effort into her **Edible Schoolyard Project**. The stated mission is to "transform the health and values of every student by building and sharing a food curriculum for the school system: http://www.edible schoolyard.org/.

Urban Beekeeping

New York–based *Borough Bees* is an excellent blog by an experienced urban beekeeper. In addition to regular updates, "Beekeeping 101" is a good primer for any would-be urban beekeeper: http://www.borough bees.com/.

Search the Internet for your local beekeepers association for local links, information, and assistance on beekeeping in your city, where permitted.

Urban Chickens

There are many urban chicken blogs and resources. For starters, there is *Toronto Chickens* blog, with helpful basics on keeping urban chickens and a listing of cities where bylaws permit small flocks in the city limits: http://torontochickens.com/.

Vertical Farming

The Plant Chicago, the world's first vertical farm posts information about its progress and systems at http://www.plantchicago.com/.

The Vertical Farm is the companion website to the book of the same name. It has a gallery of designs and a blog at http://www.vertical farm.com/.

WindowFarms is a commercial business based in New York, NY, well worth mentioning. It has instructions on how to build a vertical farm in your home or apartment out of recycled water bottles and some plastic tubing: http://www.windowfarms.org/.

ONLINE RESOURCES: INTERNATIONAL

Capital Growth is London's municipally supported urban-agriculture development and resource organization that aims to create 2,012 new community food-growing spaces across London by the end of 2012: http://www.capitalgrowth.org/.

The **London Beekeepers Association** represents urban beekeepers in the central London area: http://www.lbka.org.uk/.

The **National Society of Allotment & Leisure Gardeners** is the national representative body for the allotment movement in the United Kingdom: http://www.nsalg.org.uk.

One of **Slow Food International's** priority projects is its **A Thousand Gardens in Africa**, which will aim to set up school, community, and municipal gardens in every African country: http://www.slowfood .com/education/pagine/eng/pagina.lasso?-id_pg=24.

The **Soil Association** promotes ecologically sound farming and sustainable food through its educational campaigns and community programs throughout the United Kingdom: http://www.soilassociation.org/.

Sustain is the London-based alliance of over one hundred different groups in the United Kingdom. Sustain does a tremendous amount of work in urban food security and urban agriculture. The website is a great resource for information and educational material: http://sustain web.org/.

Urban Bees is the London-based website of urban beekeepers and beekeeping instructors Brian McCallum and Alison Benjamin: http://www.urbanbees.co.uk/.

Vertical Veg is a London-based social enterprise that inspires and supports people with very limited space who wish to grow lots of food: http://www.verticalveg.org.uk/.

*   *   *

I will continue to post urban-agriculture stories and chronicles from my urban-agriculture interviews and adventures at my Foodgirl website at http://foodgirl.ca. Join me on Facebook at https://www.facebook .com/FoodandtheCity, and follow me on Twitter at http://twitter .com/jennifer_ck.

NOTES

INTRODUCTION

1. The statistic that the average grocery store item was traveling 1,500 miles (2,414 kilometers), farm to consumer, is a rounding-off of the figure of 1,518 miles (2,442 kilometers) that fresh produce was found to be traveling, on average, in the 2001 "Food, Fuel, and Freeways: An Iowa Perspective on How Far Food Travels, Fuel Usage, and Greenhouse Gas Emissions" report from the Leopold Center for Sustainable Agriculture at Iowa State University. The lead researcher on the report is Rich Pirog.

2. Jerome Taylor, "How the Rising Price of Corn Made Mexicans Take to the Streets," *Independent*, June 23, 2007, http://www.independent.co.uk/news/world/americas/how-the-rising-price-of-corn-made-mexicans-take-to-streets-454260.html (accessed August 22, 2011).

3. Sarah Bridge, "Pasta Strike 'Shocks' Italy," *Guardian*, September 13, 2007, http://www.guardian.co.uk/world/2007/sep/13/italy (accessed August 22, 2011).

4. "Argentines Launch Tomato Boycott," BBC News, October 8, 2007, http://news.bbc.co.uk/2/hi/americas/7034152.stm (accessed August 22, 2011).

5. "World Population Prospects," United Nations, Department of Economic and Social Affairs, Population Division, Population Estimates and Projects Section, 2008 Revision, figures updated November 10, 2010, http://esa.un.org/unpd/wpp/index.htm (accessed August 22, 2011).

6. "Trade, Foreign Policy, Diplomacy, and Health," World Health Organization, http://www.who.int/trade/glossary/story028/en/ (accessed November 8, 2011).

7. "Food Security in the United States: Key Statistics and Graphics," Eco-

nomic Research Service/USDA's Briefing Room, http://www.ers.usda.gov/briefing/foodsecurity/stats_graphs.htm (accessed August 22, 2011).

8. Ibid.

9. Mark Nord et al., "Household Food Security in the United States, 2009," ERS/USDA, November 2010, p. 1. Available for download at http://www.ers.usda.gov/Briefing/FoodSecurity/stats_graphs.htm.

10. Ibid., p. 2.

11. Erik Millstone and Tim Lang, *The Atlas of Food: Who Eats What, Where, and Why* (Berkeley: University of California Press, 2008), p. 54.

12. Ibid.

CHAPTER 1. THE FACADE OF THE MODERN GROCERY STORE

1. "Supermarket Facts, Industry Overview 2010," Food Marketing Institute, http://www.fmi.org/facts_figs/?fuseaction=superfact (accessed August 22, 2011).

2. Raj Patel, *Stuffed and Starved: The Hidden Battle for the World's Food System* (Toronto, ON: Harper Perennial, 2009), p. 90.

3. World Retail Hall of Fame: Clarence Saunders, http://www.worldretailcongress.com/hall-of-fame-member-detail.cfm?id=180 (accessed May 24, 2011).

4. Ibid.

5. Bryan Walsh, "Getting Real about the High Price of Cheap Food," *Time*, August 21, 2009, http://www.time.com/time/health/article/0,8599,1917458,00.html (accessed May 24, 2011).

6. "Supermarket Facts, Industry Overview 2010," Food Marketing Institute, http://www.fmi.org/facts_figs/?fuseaction=superfact (accessed August 22, 2011). It is worth noting that this number is significantly lower than the average number of items carried in a grocery store for 2009, which was 48,750.

7. USDA ERS Food CPI and Expenditures, Table 7, http://www.ers.usda.gov/Briefing/CPIFoodAndExpenditures/Data/Expenditures_tables/table7.htm (accessed July 18, 2011).

8. Erik Millstone and Tim Lang, *The Atlas of Food: Who Eats What, Where, and Why* (Berkeley: University of California Press, 2008), p. 58.

9. Ibid.

10. "Who Will Feed Us? Questions for the Food and Climate Crises," ETC Group Communiqué, issue 102 (November 2010), p. 12. This report is also found online at http://www.etcgroup.org.

11. Ibid., p. 10.

12. Ibid., p. 12.

13. For a current, in-depth look into modern industrial agriculture, I recommend reading Barry Estabrook's *Tomatoland: How Modern Industrial Agriculture Destroyed Our Most Alluring Fruit* (Kansas City: Andrews McMeel, 2011).

14. This statistic comes from the "Fact Sheet," p. 8, which can be downloaded at http://www.foodincmovie.com/, the online tie-in site for producer/director Robert Kenner's *Food, Inc.* (New York: Magnolia Pictures, 2009).

15. Rosie Boycott, "Nine Meals from Anarchy—How Britain Is Facing a Very Real Food Crisis," *Daily Mail*, June 7, 2008, http://www.dailymail.co.uk/news/article-1024833/Nine-meals-anarchy—Britain-facing-real-food-crisis.html (accessed May 24, 2011).

16. "Supermarket Facts, Industry Overview 2010," Food Marketing Institute, http://www.fmi.org/facts_figs/?fuseaction=superfact (accessed August 22, 2011).

17. Michael Pollan, *The Omnivore's Dilemma: A Natural History of Four Meals* (Toronto, ON: Penguin, 2007), p. 11.

CHAPTER 2. INDUSTRIAL FOOD

1. *Food, Inc.*, produced and directed by Robert Kenner (New York: Magnolia Pictures, 2009).

2. Charles Lathrop Pack, *The War Garden Victorious: Its Wartime Need and Its Economic Value in Peace* (Philadelphia: J. B. Lippincott Company, 1919), p. 15.

3. Victory Gardens, Prelinger Archives, Library of Congress, Motion Picture, Broadcasting, and Recorded Sound Division, http://www.archive.org/details/victory_garden (accessed November 8, 2001).

4. T. H. Everett and Edgar J. Clissold, *Victory Backyard Gardens: Simple Rules for Growing Your Own Vegetables* (Racine, WI: Whitman Publishing Company, 1942), p. 5.

5. Michael Pollan, "Farmer in Chief," *New York Times Magazine*, October 12, 2008, http://www.nytimes.com/2008/10/12/magazine/12policy-t.html (accessed April 13, 2009).

6. Quote reproduced in Carol Ferguson and Margaret Fraser's *A Century of Canadian Home Cooking: 1900 through the '90s* (Winnipeg, MB: Prentice Hall Canada, 1992), p. 118.

7. Pollan, "Farmer in Chief."

8. Joseph Chamie, "World Population in the 21st Century," paper presented at the Twenty-Fourth IUSSP General Population Conference in Salvador, Brazil, August 18–24, 2001, p. 4. Chamie was director of the United Nations Population Division.

9. Ibid.

10. "The Nobel Prize in Chemistry 1918: Fritz Haber," NobelPrize.org, http://nobelprize.org/nobel_prizes/chemistry/laureates/1918/haber-bio.html (accessed August 22, 2011).

11. Wayne Roberts, *The No-Nonsense Guide to World Food* (Oxford, UK: New Internationalist Publications, 2008), p. 31.

12. "Population 7 Billion," *National Geographic*, January 2011, p. 50.

13. "Poverty, Population and Development," United Nations Population Fund 2008 Annual Report, UNFPA.org, http://www.unfpa.org/about/report/2008/en/ch4.html (accessed August 22, 2011).

14. "Population 7 Billion," p. 35 (flap).

15. Parag Khanna, "Beyond City Limits," *Foreign Policy* (September/October 2010), http://www.foreignpolicy.com/articles/2010/08/16/beyond_city_limits?page=full (accessed August 22, 2011).

16. "Population 7 Billion," p. 68.

17. Erik Millstone and Tim Lang, *The Atlas of Food: Who Eats What, Where, and Why* (Berkeley: University of California Press, 2008), p. 44.

18. *Food, Inc.*

19. ISAAA Brief 42-2010: Highlights of "Global Status of Commercialized Biotech/GM Crops: 2010," International Service for the Acquisition of Agri-biotech Applications, http://www.isaaa.org/resources/publications/briefs/42/highlights/ (accessed November 8, 2011).

20. Ibid.

21. William Neuman and Andrew Pollack, "Farmers Cope with Roundup-Resistant Weeds," *New York Times*, May 3, 2010, http://www.nytimes.com/2010/05/04/business/energy-environment/04weed.html (accessed August 22, 2011).

22. "Tracking the Trend toward Market Concentration: The Case of the Agricultural Input Industry," United Nations Conference on Trade and Development (UNCTAD), April 20, 2006, p. 8. This document can be read online at http://www.unctad.org/en/docs/ditccom200516_en.pdf.

23. Ibid., p. 1.

24. Eric Holt-Giménez, "Policy Brief No. 16, The World Food Crisis—What's Behind It and What We Can Do about It," Food First Institute for Food and Development Policy, October 2008, p. 6.

25. Jonathan Watts, "Field of Tears," *Guardian*, September 16, 2003, http://www.guardian.co.uk/world/2003/sep/16/northkorea.wto (accessed August 22, 2011).

26. George Lerner, "Activist: Farmer Suicides in India Linked to Debt, Globalization," CNN World, January 5, 2010, http://articles.cnn.com/2010-01-05/world/india.farmer.suicides_1_farmer-suicides-andhra-pradesh-vandana-shiva?_s=PM:WORLD (accessed August 22, 2011).

27. "Percy Schmeiser's Battle," CBC News Online, May 21, 2004, http://www.cbc.ca/news/background/genetics_modification/percyschmeiser.html (accessed August 22, 2011).

28. For more information on Monsanto's history of corporate bully tactics against small farmers and agricultural producers, I recommend Donald L. Barlett and James B. Steele's "Monsanto's Harvest of Fear," published in *Vanity Fair* magazine, May 2008. The article is available online as well at http://www.vanityfair.com/politics/features/2008/05/monsanto200805.

CHAPTER 3. INDUSTRIAL EATERS

1. Peter Menzel and Faith D'Aluisio, *Hungry Planet: What the World Eats* (Napa, CA: Material World Press; Berkeley, CA: Ten Speed Press, 2005), pp. 270–71.

2. Ibid., pp. 266–67.

3. Ibid., pp. 260–61.

4. Christopher Cook, "The New Farm Crisis," *Baltimore Chronicle*,

December 1, 1999, http://baltimorechronicle.com/farms_dec99.html (accessed May 24, 2011).

5. Rich Pirog et al., "Food, Fuel, and Freeways: An Iowa Perspective on How Far Food Travels, Fuel Usage, and Greenhouse Gas Emissions," Leopold Center for Sustainable Agriculture, Iowa State University, 2001, p. 1.

6. Ibid.

7. Ibid., p. 6.

8. The figures from this paragraph are sourced from ibid., pp. 6–7.

9. Ibid., p. 7

10. Ibid., p. 1.

11. Ibid., p. 11

12. *Food, Inc.*, produced and directed by Robert Kenner (New York: Magnolia Pictures, 2009), press kit, p. 8. The press kit is available for download at http://www.foodincmovie.com/about-the-film.php.

13. "2009 Summary Report on Antimicrobials Sold or Distributed for Use in Food-Producing Animals," US Food and Drug Administration, December 9, 2010, http://www.fda.gov/ForIndustry/UserFees/AnimalDrugUserFeeAct ADUFA/ucm236149.htm (accessed May 24, 2011).

14. Bryan Walsh, "Getting Real about the High Price of Cheap Food," *Time*, August 21, 2009, http://www.time.com/time/health/article/0,8599,1917458,00 .html (accessed January 12, 2010).

15. Erik Millstone and Tim Lang, *The Atlas of Food: Who Eats What, Where, and Why* (Berkeley: University of California Press, 2008), p. 38.

16. Ibid., p. 113.

17. Ibid., p. 119.

18. The figures in this sentence and the remaining paragraph are sourced from ibid., p. 38.

19. This figure of the one and a half acres of rain forest destruction per second comes from the Rainforest Foundation's website at http://www.rainforest foundation.org.

20. Lester Brown, "Corn for Cars: Will Biofuels Starve the Developing World?" *Der Spiegel* online, April 27, 2007, http://www.spiegel.de/international/ world/0,1518,479940,00.html (accessed May 23, 2010).

21. Ibid.

22. Eric Holt-Giménez, "Policy Brief No. 16: The World Food Crisis—What's behind It and What We Can Do about It," Food First Institute for Food and Development Policy, October 2008, p. 3. Holt-Giménez refers to a leaked World Bank Report that pegged the biofuel industry as causing a 75 percent rise in world grain prices.

23. *Food, Inc.*, press kit, p. 8.

24. Millstone and Lang, *The Atlas of Food*, p. 113.

25. *Food, Inc.*, press kit, p. 9.

26. "Obesity and Overweight," Centers for Disease Control and Prevention, http://www.cdc.gov/nchs/fastats/overwt.htm (accessed May 24, 2011).

27. "Food CPI and Expenditures: Table 7," United States Department of Agriculture Economic Research Service, http://www.ers.usda.gov/Briefing/CPI FoodAndExpenditures/Data/Expenditures_tables/table7.htm (accessed May 24, 2011).

28. Raj Patel, *The Value of Nothing: How to Reshape Market Society and Redefine Democracy* (New York: Picador, 2010).

29. Bryan Walsh, "Getting Real about the High Price of Cheap Food," *Time*, August 21, 2009, http://www.time.com/time/health/article/0,8599,1917458,00 .html.

30. This figure is from Raj Patel, interview available at http://www.youtube .com and by searching online for key words "Raj Patel" and "The Real Cost of a Hamburger."

31. "Obesity in Canada: Snapshot," Public Health Agency of Canada, http://www.phac-aspc.gc.ca/publicat/2009/oc/index-eng.php (accessed May 24, 2011).

CHAPTER 4. A WORLD IN FOOD CRISIS

1. Jeff Rubin, *Why Your World Is about to Get a Whole Lot Smaller* (Toronto, ON: Vintage Canada Editions, 2010), pp. 25–26.

2. "Number of World's Hungry to Top 1 Billion this Year—UN Food Agency," UN News Centre, June 19, 2009, http://www.un.org/apps/news/story .asp?NewsID=31197 (accessed April 24, 2011).

3. Eric Holt-Giménez, "Policy Brief No. 16: The World Food Crisis—What's behind It and What We Can Do about It," Food First Institute for Food and Development Policy, October 2008, p. 1.

4. Ibid., pp. 2–3.

5. Ibid., p. 2.

6. Mark Nord et al., "Household Food Security in the United States, 2009," *USDA Economic Research Service Report Summary*, November 2010, p. 1, is the source for the 15 percent of food-insecure citizens in the United States, rounded up by the author from 14.7 percent (2009) and 14.6 percent (2008). The other statistics are from "Food Security in the United States: Key Statistics and Graphics," also by the USDA ERS, http://www.ers.usda.gov/Briefing/FoodSecurity/stats_graphs.htm (accessed February 7, 2011).

7. "World Food Prices at Fresh High, Says UN," BBC News, January 5, 2011, http://www.bbc.co.uk/news/business-12119539 (accessed February 4, 2011).

8. Ibid.

9. James Melik, "Australia's Floods Disrupt Commodity Supplies," BBC News, January 4, 2011, http://www.bbc.co.uk/news/business-12111175 (accessed February 4, 2011).

10. In 2007, on p. 2 of its "Fourth Assessment Report," the Intergovernmental Panel on Climate Change concluded that "[w]arming of the climate system is unequivocal." Furthermore, the report concluded that human activity is responsible for a large percentage of the global warming the Earth has experienced since 1950."

11. Gwynne Dyer, "The Future of Food Riots," January 9, 2011, http://gwynnedyer.com/2011/the-future-of-food-riots/ (accessed February 4, 2011).

12. Ibid.

13. Holt-Giménez, "Policy Brief No. 16: The World Food Crisis," p. 6.

14. For a thorough read on disaster capitalism, read Naomi Klein's *The Shock Doctrine: The Rise of Disaster Capitalism* (New York: Picador, 2008).

15. After ninety-nine years of intensive monocropping of corn and palm, Madagascar would be left with utterly unfarmable land.

16. John Vidal, "How Food and Water Are Driving a 21st-Century African Landgrab," *Observer*, Guardian News, March 7, 2010, http://www.guardian.co.uk/environment/2010/mar/07/food-water-africa-land-grab (accessed August 28, 2010).

17. Figures and research for this and the two following paragraphs on the African landgrab draw from ibid., and from Nancy Macdonald's "What's the New Global Source for Fresh, Shiny Produce?" *Maclean's*, August 19, 2010, http://www2.macleans.ca/2010/08/19/out-of-africa/2/.

18. Marie-Béatrice Baudet and Laetitia Clavreul, "The Growing Lust for Agricultural Lands," English transl. Leslie Thatcher, *Le Monde*, April 14, 2009.

19. "Our Vision and Mission," Walton International, http://www.walton international.com/wigi/company-overview (accessed April 26, 2011).

20. *Food, Inc.*, produced and directed by Robert Kenner (New York: Magnolia Pictures, 2009), press kit, p. 8. The press kit is available for download at http://www .foodincmovie.com/about-the-film.php.

21. Evan D. G. Fraser and Andrew Rimas, *Empires of Food: Feast, Famine, and the Rise and Fall of Civilizations* (Toronto, ON: Free Press, 2010), p. 245.

22. Vaclav Smil, *Energy Myths and Realities: Bringing Science to the Energy Policy Debate* (Washington, DC: AEI Press, 2010).

23. Erik Millstone and Tim Lang, *The Atlas of Food: Who Eats What, Where, and Why* (Berkeley: University of California Press, 2008), p. 24.

24. Ibid.

25. Vidal, "How Food and Water Are Driving a 21st-Century African Landgrab."

26. Baudet and Clavreul, "The Growing Lust for Agricultural Lands."

27. Michael Pollan, "Farmer in Chief," *New York Times Magazine*, October 12, 2008, http://www.nytimes.com/2008/10/12/magazine/12policy-t.html (accessed April 13, 2009).

28. Michael Barclay, "Indian Food Prices Hit a Major Spike," *Maclean's*, January 18, 2010, p. 27.

29. Tom Gjelten, "The Impact of Rising Food Prices on Arab Unrest," NPR Morning Edition, February 18, 2011, http://www.npr.org/2011/02/18/1338528 10/the-impact-of-rising-food-prices-on-arab-unrest (accessed August 22, 2011).

CHAPTER 5. THE NEW FOOD MOVEMENT AND THE RISE OF URBAN AGRICULTURE

1. Tim Lang, personal e-mail exchange with the author, October 21, 2010. This exchange included a copy of the article he wrote about this 1992 television segment and its effect, "Locale/Global (Food Miles)," *Slow Food*, no. 19 (May 2006): 94–97.

2. From what I could research, the first municipal food-policy council was established in 1982, in Knoxville, Tennessee. This was certainly the first city food council in North America.

3. Lang, "Locale/Global (Food Miles)," pp. 94–97.

4. Ibid.

5. The earliest mention of food miles in a newspaper—that is, *mainstream* newspaper—I could find was in an article in the *Independent*, a UK publication, written by Johanna Blythman on October 23, 1993, in the paper's Saturday food section. The article, "Eat Local and Sever the Food Chains," begins with a brief definition of food miles as "the distance food has travelled from its point of origin to you." Blythman also reports that food in the average European shopping cart has traveled over 2,200 miles. The article can be found online at http://www .independent.co.uk/life-style/food-and-drink/food—drink-eat-local-and-sever -the-food-chains-organic-farmers-are-cutting-out-the-middlemen-and-bypassing -the-supermarkets-by-delivering-their-produce-direct-to-the-local-consumer-as -joanna-blythman-explains-1512614.html (accessed May 1, 2011).

6. An updated "Food Miles Report" is scheduled to be released by the group Sustain in 2011.

7. Carl Honoré, *In Praise of Slow: How a Worldwide Movement Is Challenging the Cult of Speed* (Toronto, ON: Vintage Canada, 2004), p. 59.

8. Steve Martinez et al., "Local Food Systems: Concepts, Impacts, and Issues," *ERS Report Summary*, US Department of Agriculture Economic Research Service, May 2010, p. 1.

9. Ibid.

10. Ibid.

11. "Our Philosophy," Slow Food, http://www.slowfood.com/ international/2/our-philosophy (accessed May 1, 2011).

12. "Manifesto for Quality," Slow Food, http://www.slowfood.com/international/2/our-philosophy (accessed May 1, 2011).

13. http://www.slowfood.com

14. "Oxford Word of the Year: Locavore," *Oxford University Press OUPblog*, November 12, 2007, http://blog.oup.com/2007/11/locavore/ (accessed December 23, 2007). While the blog post skirted a definition, it explained that "the 'locavore' movement encourages consumers to buy from farmers' markets or even to grow or pick their own food, arguing that fresh, local products are more nutritious and taste better. Locavores also shun supermarket offerings as an environmentally friendly measure, since shipping food over long distances often requires more fuel for transportation."

15. "Linking Population, Poverty, and Development," United Nations Population Fund, UNFPA.org, http://www.unfpa.org/pds/urbanization.htm (accessed August 22, 2011).

16. The big shift toward urbanization will take place in Asia and Africa, which is also where the population growth was and is taking place.

17. Carolyn Steel, *Hungry City: How Food Shapes Our Lives* (London: Vintage Books, 2009).

18. Slow Food has become an immensely influential movement. It currently has 100,000 members in 153 countries, as of May 2011. Slow Food has recently begun advocating for urban agriculture as a food security and food sovereignty measure, especially in Africa. In late 2010, Slow Food announced a major new initiative called "A Thousand Gardens in Africa." The idea is to help Africa rebuild lost community food security, help protect biodiversity of traditional food crops, and raise the profile of farming and food production in the eyes of Africa's youth. Slow Food Italy will help to fund the development of five hundred community gardens, and other Slow Food member states from around the world are partnering with projects in Africa to meet this goal.

19. Erik Millstone and Tim Lang, *The Atlas of Food: Who Eats What, Where, and Why* (Berkeley: University of California Press, 2008), p. 54.

20. Martinez et al., "Local Food Systems: Concepts, Impacts, and Issues," p. 1.

21. Vandana Shiva, *Soil Not Oil: Environmental Justice in an Age of Climate Crisis* (Cambridge, MA: South End Press, 2008), p. 38.

CHAPTER 6. PARIS: THE ROOTS OF MODERN URBAN AGRICULTURE

1. Gerald Stanhill, "An Urban Agro-Ecosystem: The Example of Nineteenth-Century Paris," *Agro-Ecosystems* 3 (1977): 269.

2. Ibid., p. 277.

3. Ibid.

4. Mary Blume, "In Praise of All Local Produce Great and Small," *New York Times*, July 21, 2001, http://www.nytimes.com/2001/07/21/style/21iht-blume_ed3_.html (accessed September 23, 2010).

5. Pascale Brevet, "Farms Flee the Cities," November 18, 2010, *Atlantic*, http://www.theatlantic.com/life/archive/2010/11/farms-flee-the-cities-can-paris-and-milan-feed-themselves/66726/ (accessed December 15, 2010).

6. The web page for the City of Paris's community gardening sites and information is http://www.paris.fr/loisirs/jardinage-vegetation/jardins-partages/p9111.

7. Jean Griffault, personal interview with the author, Paris, France, October 2, 2010. Translation from the French into English is the author's.

8. Antoine Jacobsohn (King's Vegetable Garden), personal interview with the author, Versailles, France, October 3, 2010. (Interview was conducted in English.)

9. Mairie de Paris website, http://www.paris.fr/loisirs/jardinage-vegetation/vegetation/les-vignes-de-paris/rub_8348_stand_35598_port_19375 (accessed November 10, 2010).

10. Ibid.

11. Ibid.

12. Ibid.

13. Ibid.

14. Alison Benjamin, "Fears for Crops as Shock Figures from America Show Scale of Bee Catastrophe," *Observer*, May 2, 2010, http://www.guardian.co.uk/environment/2010/may/02/food-fear-mystery-beehives-collapse (accessed August 25, 2011).

15. Zach Howard, "Researchers Seek Causes of Honeybee Colony Collapse," Reuters, March 5, 2011, http://www.reuters.com/article/2011/03/05/us-honeybee-deaths-idUSTRE7242C220110305 (accessed August 25, 2011).

16. "Questions and Answers: Colony Collapse Disorder," United States Department of Agriculture, Agriculture Research Service, last modified December 17, 2010, http://www.ars.usda.gov/News/docs.htm?docid=15572 (accessed August 25, 2011).

17. Hugh Schofield, "Paris Fast Becoming Queen Bee of the Urban Apiary World," BBC News, August 14, 2010, http://www.bbc.co.uk/news/world -europe-10942618 (accessed February 11, 2011).

18. http://www.aeroportsdeparis.fr/ADP/fr-FR/Passagers/Actualites/ Groupe-Aeroports-de-paris/Charte-abeille.htm (accessed August 23, 2011).

19. Tim Hayward, "My Bee Eats Because I'm a Londoner," video report for the *Guardian*, October 9, 2009, http://www.guardian.co.uk/lifeandstyle/wordof mouth/2009/oct/09/food-and-drink?intcmp=239 (accessed February 11, 2011).

20. David Garcelon, personal interview with the author, Toronto, Ontario, Canada, October 30, 2009.

Chapter 7. London: Capital Growth

1. Erik Millstone and Tim Lang, *The Atlas of Food: Who Eats What, Where, and Why* (Berkeley: University of California Press, 2008), p. 55.

2. Ibid.

3. "Nation of Gardeners: Russians Go Crazy about Organic Lifestyle," *Russia Today*, television video report, September 6, 2010, http://rt.com/news/ prime-time/russia-rural-urban-gardening/ (accessed October 4, 2010).

4. "Einstein Scolded for Not Weeding His Allotment—1922," City-farmer.info, December 7, 2008, http://www.cityfarmer.info/2008/12/07/ einstein-scolded-for-not-weeding-his-allotment-1922/ (accessed August 24, 2011).

5. Carolyn Steel, *Hungry City: How Food Shapes Our Lives* (London: Vintage Books, 2009), p. 5.

6. Ibid., p. 246.

7. Ibid., p. 261.

8. Rosie Boycott, "Nine Meals from Anarchy—How Britain Is Facing a Very Real Food Crisis," *Daily Mail*, June 7, 2008, at http://www.dailymail.co.uk/

news/article-1024833/Nine-meals-anarchy—Britain-facing-real-food-crisis.html (accessed August 24, 2011).

9. Tim Child (manager of St. Werburgh's City Farm), personal interview with the author, Bristol, United Kingdom, September 29, 2010.

10. Ibid.

11. Figures based on Allotment Gardens: Food and Health from NSALG website, http://www.nsalg.org.uk/.

12. T. H. Everett and Edgar J. Clisshold, *Victory Backyard Gardens: Simple Rules for Growing Your Own Vegetables with Simple Rules and Charts* (Racine, WI: Whitman Publishing Company, 1942), p. 3.

13. Figure sourced from personal website of writer/gardener Brian King, a thirty-year allotment gardener and historian of allotment gardening, at http://www.bkthisandthat.org.uk/ShortHistoryOfAllotmentshtml.html.

14. Press release, National Allotment Week 2010, National Society of Allotment & Leisure Gardeners Limited, August 2010.

15. Sean Coughlan, "Can You Dig It?" BBC News Magazine, August 11, 2006, http://news.bbc.co.uk/2/hi/uk_news/magazine/4776325.stm (accessed August 24, 2011).

16. Mark Ridsdill Smith, personal interview with the author, London, United Kingdom, September 30, 2010.

17. Monthly food value tallies are available at Ridsdill Smith's blog at http://www.verticalveg.org.uk/.

18. Matthew Weaver, "Squirrel Meat Flies off of Supermarket's Shelves," *Guardian*, July 29, 2010, http://www.guardian.co.uk/environment/2010/jul/29/squirrel-meat-supermarket (accessed August 25, 2011).

19. This claim appeared on Thorton's Budgens' website, http://www.thorntonsbudgens.com/social-environment/food-from-the-sky (accessed January 20, 2011).

20. Azul-Valérie Thomé (cocreator and project leader, Food from the Sky), personal interview with the author, London, United Kingdom, September 30, 2010.

21. "Compost Centre," Londonwaste.co.uk, http://www.londonwaste.co.uk/environment-and-society/compost-centre/ (accessed August 30, 2011).

22. Azul-Valérie Thomé, personal interview.

23. Alex Smith (founder and managing director, Alara Wholefoods), personal interview with the author, London, United Kingdom, September 28, 2010.

24. Ibid.

25. Ibid.

26. Mark Prigg, "Ideal Spot for Chateau King's Cross," *London Evening Standard*, March 16, 2009, http://www.thisislondon.co.uk/standard/article-236625 23-ideal-spot-for-chteau-kings-cross.do (accessed February 12, 2010.)

27. The information on Alara Vineyard came from my personal interview with Alex Smith, London, United Kingdom, September 28, 2010.

28. Lucy Baron Thomson (Urban Wine Company), telephone interview with the author, London, United Kingdom, May 31, 2010.

29. "Heat Island Effect," United States Environmental Protection Agency, http://www.epa.gov/heatisland/ (accessed August 25, 2011).

30. Ibid.

31. The population of Greater London, the city's administrative catchment, comes from "2010 Mid-Year Population Estimates" (revised November 2011) from the Greater London Authority's website at http://www.london.gov.uk/who -runs-london/mayor/publications/society/facts-and-figures/population (accessed August 29, 2011).

32. "High Temperatures and the Urban Island Effect," London Climate Change Partnership, http://www.london.gov.uk/lccp/ourclimate/overheating.jsp (accessed August 25, 2011).

33. This equivalence was arrived at by comparing average daily summer temperatures in London with those of the main wine-growing regions in northern France and Germany.

34. Harvest statistics can be found at http://www.urbanwineco.com/the experience.html.

35. Also sourced from the Urban Wine Company's website at http:// www.urbanwineco.com.

36. http://www.capitalgrowth.org/.

37. Bert Dutka (King's Cross Skip Garden), personal interviews with the author, London, United Kingdom, September 28, 2010.

38. Paul Richens (King's Cross Skip Garden), personal interviews with the author, London, United Kingdom, September 28, 2010.

39. Ibid.

40. The official website for the London Organizing Committee of the Olympic and Paralympic Games 2012 is http://www.london2012.com/.

41. The LOCOG document "For Starters" can be downloaded at http://www.london2012.com/documents/locog-publications/food-vision.pdf.

CHAPTER 8. SOUTHERN CALIFORNIA AND LOS ANGELES: A TALE OF TWO FARMS

1. G. Scott Thomas, "The Biggest US Metro Areas in 2025," MSNBC, June 7, 2009, http://www.msnbc.msn.com/id/31130897/ns/business-local_business/t/biggest-us-metro-areas/#.TmEix5h_wSk (accessed August 19, 2011). Note that the 12.8 million figure for the LA metropolitan area came from actual population figures for 2005, as quoted in this article.

2. Alexa Delwiche, *The Good Food for All Agenda: Creating a New Regional Foodshed for Los Angeles* (Los Angeles Food Policy Task Force, September 2010), p. 16.

3. Jonathan Gold, foreword, in Delwiche, *The Good Food for All Agenda*, p. 8.

4. Michael Ableman, *On Good Land: The Autobiography of an Urban Farm* (San Francisco: Chronicle Books, 1998), p. 33.

5. Michael Ableman, "Feeding the Future," lecture given on November 17, 2010, at the Clarke Forum for Contemporary Issues at Dickinson College, Pennsylvania. The video of this lecture is available online at http://clarke.dickinson.edu/michael-ableman/.

6. Ableman, *On Good Land*, p. 134.

7. "Farming on the Edge," American Farmland Trust, http://www.farmland.org/resources/fote/default.asp (accessed May 10, 2011).

8. Robert Gottlieb, "Showdown at South Central Farm," *Next American City*, Winter 2007, http://americancity.org/magazine/article/showdown-at-south-central-farm-gottlieb/ (accessed July 14, 2010).

9. Ibid.

10. Tezozomoc, personal interview with the author, Van Nuys, California, June 3, 2010.

11. Maxine Waters, speech given at the opening of the new South Central Farm, attended by author, June 12, 2010, Buttonwillow, California.

12. John Quigley, personal interview with the author, Buttonwillow, California, June 12, 2010.

13. Ibid.

Chapter 9. Vancouver: Canada's Left Coast

1. "Vancouver 2020—A Bright Green Future" is a downloadable document that can be found online at http://vancouver.ca/greencapital/index.htm (accessed November 10, 2011). The figures in this paragraph are all from this source.

2. "Survey and Market Assessment for Backyard Composting and Grasscycling," Mustel Group Market Research for the City of Vancouver, May 2007, http://vancouver.ca/fs/bid/bidopp/RFP/documents/PS10088AttachmentA -CityofVancouverBackyardComposting.pdf (accessed February 11, 2011).

3. Michael Levenston, personal interview with the author, Vancouver, British Columbia, July 19, 2010.

4. City Farmer's original website (http://www.cityfarmer.org/), remains online as an archive of materials posted between its 1994 launch and January 1, 2008. This is a great resource for anyone looking to trace the timeline of various trends in urban agriculture. City Farmer's new site is located at http://www.cityfarmer.info/.

5. David Tracey, personal interview with the author, Vancouver, British Columbia, July 20, 2010.

6. Seann Dory (SOLEfood farm manager), personal interview with the author, Vancouver, British Columbia, July 21, 2010.

7. Quote pulled from http://vancouver.ca/commsvcs/socialplanning/ initiatives/foodpolicy/policy/history.htm (accessed May 14, 2011).

8. Herb Barbolet, personal interview with the author, Vancouver, British Columbia, July 21, 2010.

9. Curtis Stone (owner/operator/farmer, Green City Acres), personal interview with the author, Kelowna, British Columbia, September 12, 2010.

10. "What's SPIN," http://www.spinfarming.com/whatsSpin/ (accessed February 11, 2011).

11. Green City Acres' website is at http://www.greencityacres.com/.

12. SPIN Farming's official website is http://www.spinfarming.com/.

13. Wally Satzewich, telephone interview with the author, January 20, 2011.

14. Roxanne Christensen, telephone interview with the author, January 21, 2011.

15. Wally Satzewich, telephone interview with the author, January 20, 2011.

Chapter 10. Toronto: Cabbagetown 2.0

1. The figures for the percentage of Toronto homes where some food is grown in the yard comes from Erik Millstone and Tim Lang, *The Atlas of Food: Who Eats What, Where, and Why* (Berkeley: University of California Press, 2008), p. 54. The statistics on the number of community gardens in Toronto was sourced from Theresa Boyle, "City Sees Boom in Urban Gardening," *Toronto Star*, May 22, 2011. Article can be found online at http://www.thestar.com/news/article/995391—city-sees-boom-in-urban-gardening (accessed May 26, 2011).

2. Boyle, "City Sees Boom in Urban Gardening."

3. Debbie Field, personal interview with the author, Toronto, Ontario, October 8, 2010.

4. "Toronto's Racial Diversity," City of Toronto website, http://www.toronto.ca/toronto_facts/diversity.htm (accessed August 24, 2011).

5. Ibid.

6. Field refers to statistical work done by Toronto food writer Darcy Higgins in her personal interview of October 8, 2010. This figure can also be found at http://pushfoodforward.com/category/topics/Farmers%27%20markets?page=6 (accessed August 24, 2011).

7. "Greenbelt Protection," Ontario Ministry of Municipal Affairs and Housing, http://www.mah.gov.on.ca/Page187.aspx (accessed March 1, 2011).

8. The Stop Community Food Centre's website is http://www.thestop.org/.

9. Wayne Roberts, personal interview with the author, Toronto, Ontario, October 7, 2010.

10. John H., personal interview with the author, Toronto, Ontario, October 7, 2010.

11. This figure, accurate as of May 26, 2011, was sourced from http://www.torontochickens.com/Toronto_Chickens/Where_are_chickens_legal.html.

12. Lorraine Johnson, *City Farmer: Adventures in Urban Food Growing* (Vancouver, BC: Greystone Books, 2010).

13. Lorraine Johnson, personal interview with the author, Toronto, Ontario, October 8, 2010.

14. Laura Reinsborough, telephone interview with the author, April 27, 2011.

15. "About," Not Far from the Tree, http://www.notfarfromthetree.org/about (accessed April 27, 2011).

16. "Mission Statement," FoodForward's website and "fruit counter," http://foodforward.org/about/ (accessed May 27, 2011).

17. "About Us," Portland Fruit Tree Project, http://portlandfruit.org/ (accessed May 27, 2011).

18. Tim Kitchen, telephone interview with the author, January 20, 2011.

19. Darrin Nordahl, telephone interview with the author, September 8, 2010.

20. Darrin Nordahl, *Public Produce: The New Urban Agriculture*. (Washington, DC: Island Press, 2009).

21. Darrin Nordahl, "Smart Governments Grow Produce for the People," http://www.grist.org/article/food-smart-city-governments-grow-produce-for-the-people, August 5, 2010 (accessed May 27, 2011).

22. Wayne Roberts, "Eat this Recession," Alternatives online journal, October 15, 2009, http://www.alternativesjournal.ca/articles/eat-this-recession (accessed May 19, 2011).

23. *The Diane Rehm Show*, NPR, first aired October 25, 2009, http://the dianerehmshow.org (accessed September 6, 2010). (Search on the site for key words "Darrin Nordahl" and "public produce.")

Chapter 11. Milwaukee: Growing a Social Revolution

1. Elizabeth Royte, "Street Farmer," *New York Times Magazine*, July 1, 2009, http://www.nytimes.com/pages/magazine/index.html (accessed January 11, 2010).

2. "Together We Are Growing Power," Growing Power, Inc., press kit, 2011, provided by Will Allen. Growing Power's press kit is available for download at http://www.growingpower.org/assets/presskit.pdf.

3. Information on Will Allen's background and company history was compiled primarily from ibid., and from Royte, "Street Farmer."

4. Royte, "Street Farmer."

5. On January 27, 2011, the author attended "Growing Out of Hunger," a talk by Will Allen at Simon Fraser University Centre for Dialogue in Vancouver, British Columbia.

6. Growing Power's mission statement is taken from its "Together We Are Growing Power" press kit, 2011.

Chapter 12. Detroit: Praying for an Economic Revolution

1. Tim Jones, "Detroit's Outlook Falls Along with Home Prices," *Chicago Tribune*, January 29, 2009, http://articles.chicagotribune.com/2009-01-29/news/0901280800_1_home-prices-mayor-kwame-kilpatrick-outlook (accessed February 25, 2011).

2. David Whitford, "Can Farming Save Detroit?" CNNMoney.com, December 29, 2009, quoting Doug Rothwell, CEO of Business Leaders for Michigan, http://money.cnn.com/2009/12/29/news/economy/farming_detroit.fortune/ (accessed February 24, 2011).

3. Catherine Porter, "From Motown to Hoetown," *Toronto Star*, September 26, 2009, http://www.thestar.com/news/insight/article/700654 (accessed August 25, 2011).

4. "Detroit Poverty Getting Worse," hosted by Farai Chideya, NPR,

October 5, 2005, http://www.npr.org/templates/story/story.php?storyId=4955 488 (accessed August 25, 2011).

5. Daniel Okrent, "Detroit: The Death—and Possible Life—of a Great City," *Time*, September 24, 2009, http://www.time.com/time/magazine/article/ 0,9171,1926017,00.html (accessed on August 27, 2011).

6. Ibid.

7. Jones, "Detroit's Outlook Falls Along with Home Prices."

8. "State & County Quickfacts: Detroit (city), Michigan," US Census, last revised July 8, 2009, http://quickfacts.census.gov/qfd/states/26/2622000.html (accessed May 24, 2011).

9. Alethia Carr, telephone interview with the author, Detroit, Michigan, February 2, 2011.

10. Definition comes from the federal WIC website at http://www.fns.usda .gov/wic/aboutwic/ (accessed May 24, 2011).

11. "WIC-Fact-Sheet.pdf," p. 2, downloaded from http://www.fns.usda .gov/wic/aboutwic/ (accessed August 27, 2011).

12. Taja Sevelle's Urban Farming program, based in Detroit, has started farms on unused land and grows free vegetables and produce for the community surrounding the farm, http://www.urbanfarming.org.

13. Greening of Detroit web address is http://www.greeningofdetroit.com/.

14. Detroit Black Community Food Security Network, among other food -security organizations, operates the D-Town Farm. The group's web address is http://detroitblackfoodsecurity.org/.

15. This figure of 1,300 urban food gardens in Detroit was put forward by Dan Carmody, manager of Detroit's Eastern Markets, a well-known six-block farmers' market. Alethia Carr was in attendance and reported on this figure at http://www.foodandsocietyfellows.org/digest/article/detroit-business-urban-agriculture (accessed August 27, 2011).

16. "World's Largest Urban Farm Planned for the City of Detroit," Hantz Farms press release, April 23, 2009, http://www.hantzfarmsdetroit.com/ press.html (accessed December 22, 2010).

17. Shea Howell, "Detroit Counts," *Michigan Citizen*, April 3, 2011, http://michigancitizen.com/detroit-counts-p9645-1.htm (accessed May 24, 2011).

18. The panel discussion is available online at http://www.umd.umich.edu/urbanfarming.

19. John Hantz quotes are from the panel discussion "Urban Farming Summit: The Business of Urban Agriculture," University of Michigan–Dearborn, April 7, 2010. A video of the conference is available at http://www.umd.umich.edu/urbanfarming.

20. Ibid.

21. "Homestead Act 2010," City of Beatrice, Nebraska, http://www.beatrice.ne.gov/departments/city/attorney/homestead.shtml (accessed August 25, 2011).

22. Carlton Flakes, panel discussion, "Urban Farming Summit: The Business of Urban Agriculture," April 7, 2010.

23. Mike Score (president of Hantz Farms), personal interview with the author, Detroit, Michigan, February 1, 2011.

24. Ibid.

25. Ibid.

CHAPTER 13. CHICAGO: THE VERTICAL FARM

1. Dickson Despommier, *The Vertical Farm: Feeding the World in the 21st Century* (New York: Thomas Dunne Books, 2010).

2. Blake Kurasek's Living Skyscraper can be viewed at http://blakekurasek.com/thelivingskyscraper.html.

3. Vincent Callebaut's Dragonfly vertical farm can be viewed at http://vincent.callebaut.org/page1-img-dragonfly.html.

4. Despommier and Ellingsen's pyramidal vertical farm can be viewed at http://www.verticalfarm.com/designs?folder=b9aa20a4-9c6a-4983-b3ad-390c4f1fa562.

5. MVRDV's website is at http://www.mvrdv.nl.

6. The Plant's website is at http://www.plantchicago.com/.

7. Blake Davis (adjunct professor of urban agriculture at Illinois Institute of Technology), personal interview with the author, The Plant, Chicago, Illinois, January 29, 2011.

8. Personal interviews were conducted onsite at The Plant with John Edel (owner/developer/director, The Plant), Blake Davis (adjunct professor of urban

agriculture, IIT), Alex Poltorak (volunteer, The Plant), and Audrey Thibault (volunteer, The Plant), Chicago, Illinois, January 29, 2011.

9. I found several sources that reference the influence of the Chicago Stockyard's "disassembly line" on Henry Ford's idea for the automobile assembly line. He saw the efficiencies gained by giving one worker one specific task and then moving the carcasses on to the next worker. Ford reversed the process to put cars together, but the idea of worker specialization was born on the blood-soaked floors of stockyard slaughterhouses. One source, among many online, that states this is http://www.pbs.org/wgbh/amex/chicago/peopleevents/p_armour.html.

10. The backstory of how John Edel came to purchase the Peer Foods building and information on The Plant came from a personal interview with John Edel, January 29, 2011.

11. Bubbly Dynamics draws its nickname from nearby Bubbly Creek, a waterway named during the days of the stockyards and the attendant business that sprung up around the century-long livestock and slaughter industry in Chicago, where boiled waste and decaying matter made the creek appear to bubble.

12. John Edel, personal interview with the author, The Plant, Chicago, Illinois, January 29, 2011.

13. Ibid.

14. John Edel and Nathan Wyse in conversation with the author, The Plant, Chicago, Illinois, January 29, 2011.

15. John Edel, personal interview with the author, The Plant, Chicago, Illinois, January 29, 2011.

16. Blake Davis, in person interview with the author, The Plant, Chicago, Illinois, January 29, 2011.

17. Ibid.

CHAPTER 14. CUBA: URBAN AGRICULTURE ON A NATIONAL SCALE

1. Bill McKibben, "The Cuba Diet: What Will You Be Eating When the Revolution Comes?" *Harper's* magazine, April 2005, http://harpers.org/archive/2005/04/0080501 (accessed March 20, 2007).

2. Fernando Funes et al., *Sustainable Agriculture and Resistance: Transforming Food Production in Cuba* (Oakland, CA: Food First Books, 2002), p. 5.

3. Ibid., p. 37.

4. These details were compiled from personal conversations between Cuban citizens and the author, February 2007 and May 2010.

5. McKibben, "The Cuba Diet."

6. Funes et al., *Sustainable Agriculture and Resistance*, p. 5.

7. Ibid.

8. Ibid.

9. McKibben, quoting Fernando Funes in "The Cuba Diet."

10. Ibid.

11. In Spanish, this is referred to as *Período Especial en Tiempo de Paz*, or, simply, *Período Especial*.

12. Funes et al., *Sustainable Agriculture and Resistance*, p. 235.

13. Sarah Murdoch, "Future of Food: The Vegetable Gardeners of Havana," BBC Two, August 22, 2009, http://news.bbc.co.uk/2/hi/8213617.stm (accessed May 24, 2010).

14. "The Accidental Revolution, Pt. 1," *The Nature of Things*, CBC Television, 2006.

15. Living Planet Report 2006, World Wildlife Fund, http://wwf.panda.org/about_our_earth/all_publications/living_planet_report/living_planet_report_timeline/lp_2006/ (accessed August 26, 2007). (It should be noted that these are based on self-reported statistics from the countries.)

16. Personal attendance at VIII Meeting on Organic and Sustainable Agriculture international conference, in Havana, Cuba, May 11 to 14, 2010.

17. Personal visit to Little Radish *organopónico*, Ciego de Ávila, Cuba, February 5, 2007.

18. Jorge Carmenate (manager, Little Radish *organopónico*), personal interview with the author, Ciego de Ávila, Cuba, February 5, 2007.

19. Ibid.

20. The author calculated this based on the number of *organopónicos* in Ciego de Ávila and the population figures of the city.

21. Cuba has two currencies: one is the Cuban peso (CUP), an internal currency that Cuban nationals use, often referred to as "national money." State wages

are paid in CUPs, as are essential items, such as food and medicine. Cuban pesos are generally not available to tourists and have no real value outside Cuba; therefore, it is difficult and often meaningless to convert CUP prices into dollars. Cuban convertible pesos (CUC), is the currency used by tourists and by Cubans for luxury items. It is referred to as "hard currency," as it is exchangeable for US dollars on a 1:1 ratio, plus a 10 percent tax at the time of conversion.

22. Personal visit to Vivero Alamar, Havana, Cuba, May 10, 2010.

23. The author compiled this list of ration items and quantities from interviews conducted with our Cuban tour guides and translators in May 2010.

24. Personal visit to ration store, Camagüey, Cuba, February 6, 2007.

25. Clifford L. Staten, *The History of Cuba* (New York: Palgrave Macmillan, 2003), p. 6.

26. Personal visit to Agromercado 19 y B, Havana, Cuba, May 15, 2010.

27. Tito Nuñez (chef of El Romero) personal interview with the author at the eco-restaurant Las Terrazas, Havana, Cuba, May 12, 2010.

28. Ibid.

Chapter 15. Conclusion: Greening and Eating Our Cities

1. Brooklyn Grange Farm, Queens, New York, http://www.brooklyn grangefarm.com.

2. Eagle Street Rooftop Farm, Queens, New York, http://rooftopfarms.org.

3. Lufa Farms, Montreal, Quebec, https://lufa.com/en.

4. "Rooftop Farm Set to Deliver Fresh Produce to Montreal Consumers," Lufa Farms press release, March 22, 2011, http://www.cnw.ca/en/releases/archive/March2011/22/c5312.html (accessed August 27, 2011).

5. Gotham Greens, Brooklyn, NY, http://gothamgreens.com/.

6. Ibid.

7. Design a School Garden (with LAUSD) and We'll Build It !" Good Education, April 26, 2010, http://www.good.is/post/design-a-school-garden-with-lausd-and-we-ll-build-it/ (accessed August 27, 2011).

8. Ibid.

9. Terence Belford, "Developers Blue over Green Roofs," *Globe and Mail*, June 16, 2009 (updated January 10, 2010), http://www.theglobeandmail.com/report-on-business/developers-blue-over-green-roofs/article1183436/ (accessed August 26, 2011).

10. Ron Berezan, personal interview with the author, Edmonton, Alberta, April 29, 2009.

11. Ibid.

12. Ibid.

13. Figures sourced from "Intergovernmental Dialogue," United Nations Department of Economic and Social Affairs, http://www.un.org/en/development/desa/financial-crisis/government-dialogue.shtml (accessed May 31, 2011).

14. Lisa Foderaro, "Enjoy Park Scenery, City Says, But Not as a Salad," *New York Times*, July 29, 2011, http://www.nytimes.com/2011/07/30/nyregion/ new-york-moves-to-stop-foraging-in-citys-parks.html?pagewanted=allfourteen (accessed August 31, 2011).

BIBLIOGRAPHY

Ableman, Michael. *On Good Land: The Autobiography of an Urban Farm*. San Francisco, CA: Chronicle Books, 1998.

Brown, Celia Brooks. *New Urban Farmer, From Plot to Plate: A Year on the Allotment*. London: Quadrille Publishing, 2010.

Despommier, Dickson. *The Vertical Farm: Feeding the World in the 21st Century*. New York: Thomas Dunne Books, 2010.

Fraser, Evan D. G., and Andrew Rimas. *Empires of Food: Feast, Famine, and the Rise and Fall of Civilizations*. Toronto, ON: Free Press, 2010.

Funes, Fernando, Luis García, Martin Bourque, Nila Pérez, and Peter Rosset. *Sustainable Agriculture and Resistance: Transforming Food Production in Cuba*. New York: Food First Books, 2002.

Honoré, Carl. *In Praise of Slow: How a Worldwide Movement Is Challenging the Cult of Speed*. Toronto, ON: Vintage Canada, 2004.

Johnson, Lorraine. *City Farmer: Adventures in Food Growing*. Vancouver, BC: Greystone Books, 2010.

Kropotkin, Peter. *Fields, Factories and Workshops*. Montreal: Black Rose Books, 1994.

Millstone, Erik, and Tim Lang. *The Atlas of Food: Who Eats What, Where, and Why*. Berkeley: University of California Press, 2008.

Mougeot, Luc J. A., *Growing Better Cities: Urban Agriculture for Sustainable Development*. Ottawa, ON: International Development Research Centre, 2006.

Nordahl, Darrin. *Public Produce: The New Urban Agriculture*. Washington, DC: Island Press, 2009.

Nestle, Marion, *Food Politics: How the Food Industry Influences Nutrition and Health*. 2nd edition. Berkeley: University of California Press, 2007.

Palassio, Christina, and Alana Wilcox. *The Edible City: Toronto's Food from Farm to Fork*. Toronto, ON: Coach House Books, 2009.

Patel, Raj. *Stuffed and Starved: The Hidden Battle for the World's Food System*. Toronto, ON: Harper Perennial, 2009.

———. *The Value of Nothing: Why Everything Costs So Much More than We Think*. Toronto, ON: HarperCollins, 2009.

Pollan, Michael. *In Defense of Food: An Eater's Manifesto*. New York: Penguin Press, 2008.

———. *The Omnivore's Dilemma: A Natural History of Four Meals*. Toronto, ON: Penguin, 2007.

Richardson, Jill. *Recipe for America: Why Our Food System Is Broken and What We Can Do to Fix It*. Brooklyn, NY: Ig Publishing, 2009.

Roberts, Wayne. *The No-Nonsense Guide to World Food*. Oxford, UK: New Internationalist Publications, 2008.

Rubin, Jeff, *Why Your World Is about to Get a Whole Lot Smaller*. Toronto, ON: Vintage Books, 2010.

Schlosser, Eric. *Fast Food Nation: The Dark Side of the All-American Meal*. New York: Houghton Mifflin, 2001.

Staten, Clifford L., *The History of Cuba*. New York: Palgrave Macmillan, 2003.

Steel, Carolyn. *Hungry City: How Food Shapes Our Lives*. London, UK: Vintage Books, 2009.

INDEX

Photos indicated by *italicized* page numbers.